D0915983

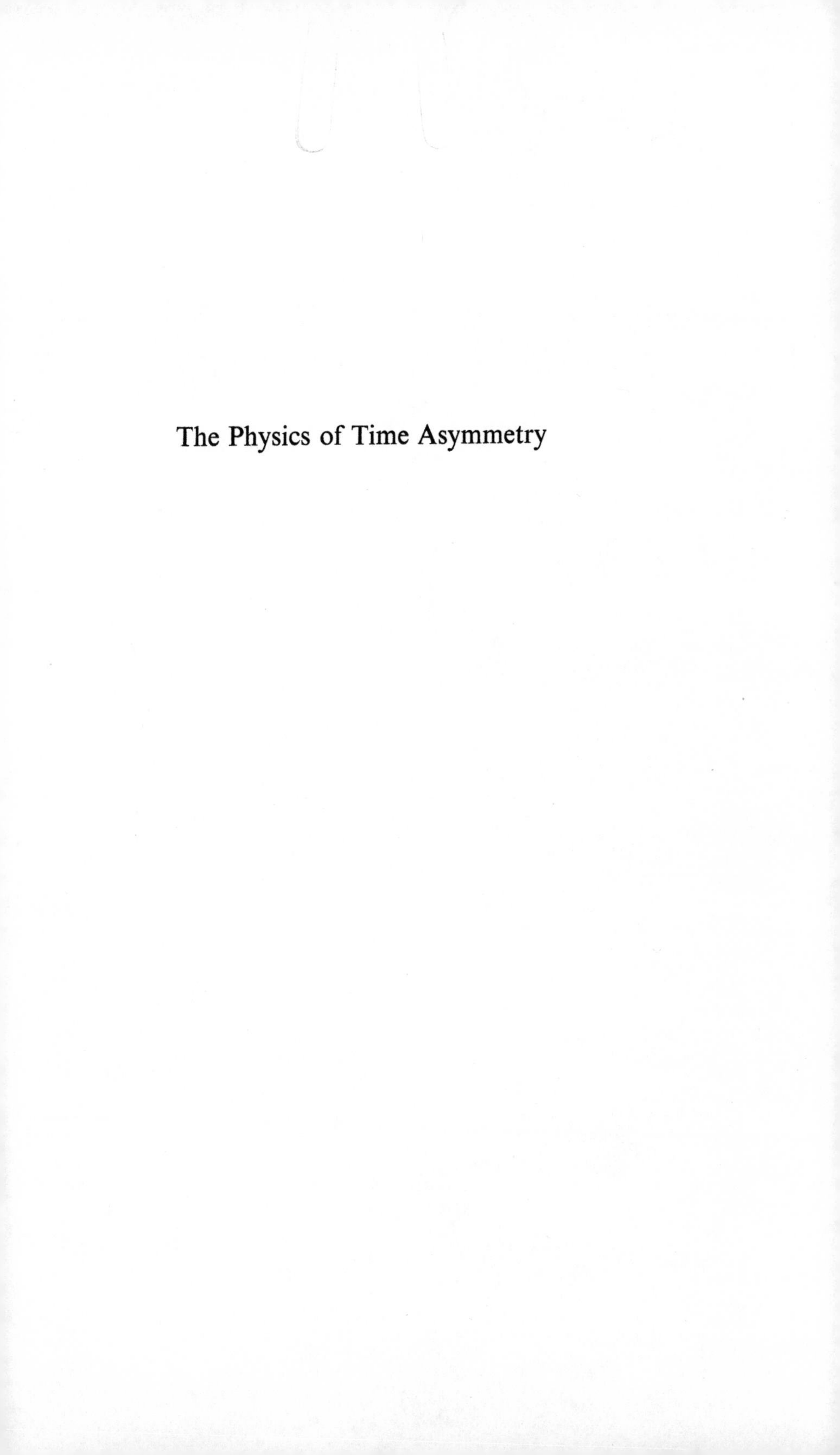

The Physics of Time Asymmetry

The Physics of Time Asymmetry

P. C. W. DAVIES

University of California Press
Berkeley and Los Angeles

Published in the UK and Commonwealth 1974 by
Surrey University Press, London.

Published in the USA 1974 by
University of California Press,
Berkeley and Los Angeles

ISBN 0-520-02825-2
Library of Congress Catalog Card Number: 74-81536

Printed in Northern Ireland at the Universities Press, Belfast.

TO SUSAN

ACKNOWLEDGEMENTS: I should like to thank Professor P. T. Landsberg, Dr D. Robinson, Dr W. C. Saslaw and Dr M. Rowan-Robinson for helpful comments and Mrs M. Woodcock for assistance with the manuscript.

Contents

Index of symbols

$\mathcal{N}$	number in ensemble 44, 161
n	neutron 105
$\boldsymbol{n}$	unit vector from source of electromagnetic disturbance 122
n	refractive index 135
n_i	chemical concentration 32, 189
n_i	occupation number of cells i in μ space 50, 63
n_i	occupation number of quantum mechanical group of states i 164
n_0	real part of refractive index 136
P	Dyson time ordering operator 180
P_ν	total probability of finding a system in the quantum mechanical group of states ν 165
p	pressure of gas 31
p	shorthand notation for $(p_1, \ldots p_{3N})$ 44
p	pressure of cosmological fluid 82
p	proton 93
$\boldsymbol{p}$	momentum vector (shorthand notation for $\{p_1, p_2, p_3\}$) 35, 177
d^3p	shorthand notation for dp_1, dp_2, dp_3 35
p^μ	momentum four-vector 177
p_i	momentum coordinates 35
Q	quantity of heat 30
q	shorthand notation for $(q_1, \ldots q_{3N})$ 44
q	deceleration parameter 85
$\boldsymbol{q}$	position vector (shorthand notation for $\{q_1, q_2, q_3\}$) 35
d^3q	shorthand notation for dq_1, dq_2, dq_3 35
q_i	position coordinates 35
R	scale factor in Robertson–Walker metric 18, 82
$R = \|\boldsymbol{R}\|$	114, 122
$\boldsymbol{R}$	vector displacement from source of wave disturbance 115, 122
ΔR	perturbation in scale factor R 107
$\mathcal{R}_{\mu\nu}, \mathcal{R}$	tensor and scalar quantities in field equations of general relativity 81
R_{ext}, R_{int}	external and internal radius of spherical absorbing shell 132
r	radial coordinate in Robertson–Walker metric 18, 82
r	radial coordinate of distant galaxy 87
r	radius of spherical shell of stars 94
r	displacement from centre of absorbing cavity 136
$\boldsymbol{r}$	position vector in space 23, 112
$\boldsymbol{r}'$	position vector of wave source 113

$\mathbf{\Lambda}$	quantity related to black hole energy 100
λ	wavelength of electromagnetic radiation 89
λ_i	parameter along world line of particle i 121
μ, ν	tensor indices 10
μ^+, μ^-	μ mesons 176
ν	effective collision frequency of ions 147
$\nu_e, \bar{\nu}_e$	electron neutrino and antineutrino respectively 105, 176
$\bar{\nu}_\mu$	muon antineutrino 176
Π	'temperature' of black hole 100
π^+, π^-	π mesons 176
P	coarse-grained density of points in Γ space 50
P	coarse-grained statistical operator 165
P_n	coarse-grained density for star n 65
P_{nn}	coarse-grained density matrix element 165
ρ	density of points in Γ space 44
ρ	density of stars in space 95
ρ	density of source of scalar wave 112
ρ	density of absorbing charged particles 136
ρ	density of absorbing bodies in general 145
ρ	density of protons 196
$\hat{\rho}$	statistical operator 161
ρ_γ	density of photons 89, 196
ρ_i	density of ions 145
$\hat{\rho}_{nm}$	density matrix element 161
$\hat{\rho}_{nm}^{(i)}, \hat{\rho}_{nm}^{(2)}$	density matrix element for first and second measurement processes, respectively 168
σ	cross section for particle scattering 38
σ	fine-grained H function 49, 162
σ	'entropy' of black hole 100
σ	cross section for absorbing bodies 145
σ	cross section for inverse bremmstrahlung 146, 186
$\boldsymbol{\sigma}$	spin vector 155
σ^1	cross section for inverse particle scattering 39
σ_T	cross section for Thomson scattering 147
τ_i	proper time of particle i 120
Φ_n	eigenfunction of measuring apparatus in state n 170
ϕ	amplitude of scalar wave 112
ϕ	electromagnetic scalar potential 122
ϕ	angular displacement from vector $\boldsymbol{U}_0$ 135
ϕ, ϕ'	quantum mechanical state vectors 154
ϕ_{ret}, ϕ_{adv}	retarded and advanced scalar wave amplitudes respectively 113, 114

$\psi_1, \psi_2, \psi_3, \bar{\psi}_2$	states of gas 58
ψ, ψ'	quantum mechanical state vectors 154
Ω	angular coordinates in Robertson–Walker metric 18, 82
Ω	angular coordinates for particle scattering 38
Ω	angular coordinates in absorbing cavity 136
Ω_A, Ω_B	regions of Γ space 56, 57
Ω, Ω'	regions of Γ space analogue 101
ω	angular frequency of electromagnetic radiation 89, 134
ω	angular frequency of oscillating charged particle 134
ω_p	plasma frequency 147
$\mathbf{\nabla}_6 \cdot$	divergence in μ space 35
$\mathbf{\nabla}_{6N} \cdot$	divergence in Γ space 44
$\mathbf{\nabla}_q$	gradient operator acting on position vector 35
$\mathbf{\nabla}_v$	gradient operator acting on velocity vector 36
∇^2	Laplacian operator 112
$\square^2$	d'Alembertian operator 121
$\vert\rangle$	ket vector 154
$\langle\vert$	bra vector 154
$\dagger$	Hermitean conjugation superscript 154
$*$	complex conjugation superscript 156
Tr	trace 161

Introduction

How is it possible to account for the difference between past and future when an examination of the laws of physics reveals only the symmetry of time? Scarcely any topic in fundamental physics can avoid running up against this problem at some stage; and yet after more than a century of speculation, occupying the attention of some of the world's greatest physicists, the question seems far from answered. It is certainly extraordinary that the explanation of such a fundamental aspect of everyday experience should remain obscure and paradoxical after consideration by people like Boltzmann, Einstein, Schrödinger, Eddington and the Ehrenfests. Over the centuries, in addition to the physicists, philosophers such as Aristotle, St Augustine and Kant have been fascinated by the problems of the nature of time.

A casual inspection of the literature creates the impression of great confusion and misunderstanding surrounding the topic. The physics of time asymmetry has never been a single well-defined subject, but more a collection of consistency problems which arise in almost all branches of physics when confronted with a choice of boundary conditions compatible with the real world. As a result, different authors have approached the subject from widely varying standpoints, employing almost non-intersecting explanation schemes and terminology for what should really be the common ground of much of twentieth century physics.

The first purpose of this book is to collect together all these essentially related strands of research, and to combine them into a single subject matter with a uniform terminology and careful attention to definitions. This in itself is sufficient to remove many of the misconceptions. Furthermore, discussions of the subject that at first seem to be disjoint, or even contradictory, are shown to be complementary components in a reasonably consistent and precise theory. This achievement is only possible if a clear distinction is made between the physical and philosophical character of the problems. Much of the prevailing confusion about time asymmetry stems from a lack of precision about what is actually being discussed. For example, whether the *nature* or *origin* of the asymmetry is the question at issue. Attempts are even made to reduce the entire problem to subjective concepts, such as language and psychology[1]. In contrast, the approach adopted here is strongly rooted in the real world, and time asymmetry is analysed from the standpoint that it is the observed phenomenon in everyday life which needs a clear explanation.

Historically, physicists have had a picture of time as something like an 'ever-rolling stream'—an all pervading motion carrying the contents of the universe irresistibly from past to future. When first introduced as a mathematical parameter into the laws of physics by Newton, great emphasis was laid on the uniformity of time and the precision of its forward flow. Newtonian time was absolute and universal, and standard clocks were supposed to agree on the rate of flow irrespective of their location or motion. This picture of time is commonplace among laymen even today.

Alongside the notion of a 'flowing stream' there developed another very deep-rooted belief about temporal structure, which few people would be prepared to deny. This is the existence of a *moving present moment*. It is, of course, incontestable that human beings are strongly aware of a *now* in their consciousness, and that this now is being progressively and steadily transported from the past towards the future. Indeed the very meaning of 'past' and 'future' are determined at a psychological level by this experience. Before the days of special relativity, it was quite acceptable to suppose that the entire universe had real existence for only one instant (now!), *the* past world having passed out of existence, *the* future world yet to come into being.

The arrival of special relativity brought about the collapse of the whole concept of an absolute, uniform time, and a universal now. Time itself became part of a new structure: four dimensional *space–time*. In this new structure, intervals of space are always associated with intervals of time, so that the concept of the 'same moment' in two different places is without absolute meaning. In the new four dimensional world picture, the entire past and future history of a system must be considered to be in existence together.

The mistake of pre-relativity physics was to identify time too closely with human experience. It must not be thought from this that relativity contradicts human experience. The bold concept of a universal now has been replaced with the more modest 'here and now', the *event*, which associates an instant of time only with a single point in space. We are still permitted our own private 'nows' as our conscious awareness apparently moves pointwise along our world line paths in space–time, uniformly maybe in our own frames of reference, but without relevance or absolute meaning to anyone else. Thus, relativity physics has shifted the moving present out from the superstructure of the universe, into the minds of human beings, where it belongs. Some authors have taken this as an opportunity to relegate the now of our consciousness to the status of pure illusion, belonging properly to the realm of psychology rather than physics. Others are reluctant to abandon the physical reality of what is perhaps the most fundamental aspect of all our experience. Nevertheless, and this fact does not seem to have been sufficiently strongly emphasized to physicists, present day physics makes no provision whatever for a flowing time, or for a moving present moment. Those who might wish to retain these concepts are obliged to propose that the mind itself participates in a novel way in some form of physical activity that is not manifest in the laboratory, a suggestion that meets with a great deal of reserve among the scientific community. Eddington has written that the acquisition of information about time occurs at two levels: through our sense organs in a fashion consistent with laboratory physics, and in addition through the 'back door' of our own minds[2]. It is from the latter source that we derive the customary notion that time 'moves'.

In the absence of an acceptable theory of the mind in physics, any discussion of physical time must necessarily exclude consideration of the now, and the apparent forward flow of time, because these are meaningless concepts within the context of ordinary space–time as it is at present understood. Although this fact is well known to modern philosophers, most physicists have continued to carry over into the *passive* space–time of physics concepts from the *active* time of psychology. It has become an almost universal practice to refer to *the direction of time* or the *arrow of time* in physics, with the implicit meaning of the direction of flow or movement of the now from past to future. It is here that perhaps the most serious misunderstanding of all has arisen, because this dubious psychological concept of 'becoming' has been so frequently muddled with the objective, legitimate physical concept of time *asymmetry*.

The two directions of time in the following sense—*towards* the past and *towards* the future—are known from common experience to be fundamentally distinguished physically. This fact is quite independent of the existence or motion of the now. For example, we *remember* the past. Moreover, this *asymmetry* with respect to the two time orientations is also readily recognized in laboratory physics. Indeed, practically all the phenomena of nature appear to be asymmetric in time. As a result of the universal nature of this asymmetry there has grown the additional misconception that it is a structural property of *time itself*. A closer inspection reveals that it is more appropriate to regard the asymmetry as a collective property of physical systems *in* (space–)time. In this sense, time as such does not possess any *intrinsic* orientation, asymmetry, movement, direction or arrow. In an empty universe such things do not exist. Time asymmetry is here taken to mean the basic fact of nature that the contents of the world possess a structural distinction between past and future facing orientations.

The pioneering work on time asymmetry was carried out by Boltzmann, who applied the laws of mechanics, together with a statistical assumption, to a molecular model of a gas. The important statement about the asymmetry is contained in the famous *H* theorem concerning the increase in entropy in isolated systems. The interpretation of this theorem was for a long time the subject of controversy, until the careful work of the Ehrenfests. In particular, the reversibility objections of Loschmidt and Zermelo appeared to indicate a contradiction between experimental thermodynamics and the time reversible laws of mechanics which Boltzmann applied to individual molecules. The resolution of this paradox followed with an appreciation that the *H* theorem alone could not account for the observed asymmetry of thermodynamic systems, but required in addition an explicitly asymmetric assumption about the way in which real systems were formed in the first place. A very lucid exposition of this was given in the posthumous book by Reichenbach, published in 1956, who made use of the concept of *branch systems* which separate off from the main environment in a low entropy state.

In spite of the longstanding successful reconciliation between the reversible laws of mechanics and the time asymmetry of familiar thermodynamic processes, the reversibility objections continue to be raised from time to time in support of unconventional explanations of time asymmetry. It is a remarkable fact that the Boltzmann–Ehrenfest–Reichenbach explanation scheme has not been generally accepted by modern physicists. One reason for this is undoubtedly the fact that the *H* theorem is often taken for more than it really is, i.e. an explanation of what is *actually observed* by macroscopic human beings, and *not* an attempt at a theoretical description of a special quality with an independent existence. Contrary to widespread belief, time asymmetry is only a *type of description*, relevant to the macroscopic world-view of the physicist, rather than an extra *physical* ingredient to be added to the laws of mechanics. The latter belief, that there is something more to time asymmetry than a description of growing disorder, probably arises from the active picture of time. Proponents of this view often object that the asymmetry described by the *H* theorem is subjective, or not 'real', based on the argument that it only arises as a result of the rejection of information about physical systems, i.e. by approximating our description of nature, say by coarse graining, or neglecting to consider the detailed complicated correlations between individual atoms. Of course, that is true. But nothing yet discovered in nature requires individual atoms to experience time asymmetry, the very essence of which is the collective quality of complex systems, like life itself. Whether the asymmetry described by the *H* theorem is regarded as 'subjective' or not is only relevant to philosophy, not physics. In truth, it is *all that is needed* to successfully account for observations of the real world.

The misleading belief that nature imposes a direction of time onto everything (in the active time sense), including individual particles[3], has led many authors to look for the origin of this mysterious 'extra ingredient'. Some have appealed to the microscopic domain and the subject of quantum mechanics[4], others to cosmology[5] or a mixture of cosmology and electrodynamics[6]. Others attempt to deny or mutilate the Hamiltonian description[3], perhaps appealing to the random interactions of the system with the world outside[7]. The complexity and obscurity of many of these attempts strongly suggests that there is no additional quality to be found.

Running parallel with all these controversies in thermodynamics and statistical mechanics have been the long arguments about the boundary conditions for retarded radiation in electrodynamics. As early as 1909 Einstein and Ritz were in disagreement about whether retardation is a necessary prerequisite for the second law of thermodynamics or vice versa. That there is certainly an additional non-thermodynamic asymmetry to wave motion has been pointed out by Popper. The only convincing attempt to explain the predominance of retarded radiation over advanced, on the basis of a detailed mathematical theory, was that presented by Wheeler and Feynman in 1945. In this so-called 'absorber theory', the thermodynamic properties of the material which eventually absorbs the radiation are used to account for its retarded nature. This theory has been much championed in recent years by some cosmologists, who have replaced the thermodynamic considerations of the absorber with cosmological ones, thereby opening up the fascinating possibility of discriminating between cosmological models on the basis of local observations of retarded radiation. Unfortunately, the status of thermodynamics in these models is very obscure, and much of the published work contains confusing and unnecessarily complicated material.

A large school of thought which believes that cosmology determines the temporal asymmetry of thermodynamics and electrodynamics has grown in the last decade, though opinions differ widely as to the closeness or order of this connection. In a now classic paper, T. Gold suggested that the expansion of the universe in some way *maintains* the thermodynamic disequilibrium of the world, in the sense that in a recontracting universe the direction of thermodynamic and electrodynamic processes would reverse. This suggestion has caught the imagination of many physicists and astronomers, even though it runs counter to the pioneering work of Tolman, whose application of thermodynamics to cosmology in the 1930s demonstrated that repeated cycles of expansion and contraction of the cosmos would in general only serve to increase the entropy of the cosmological material. In any case, it would be expected on quite general grounds that the cosmological expansion would only be relevant to local thermodynamic processes when the expansion time scale is comparable with the process relaxation times. This was indeed the case in the very early stages of the so-called 'big bang' cosmological models, and it is to this big bang that the search must be directed for an explanation of the present disequilibrium of the universe.

The importance of Tolman's contribution was the discovery that *gravitation* provides the ultimate and inexhaustible reservoir of negative entropy for the universe, and as such lies at the very foundation of the subject of time asymmetry. It is curious that only in the last few years has this thermodynamic relevance of gravitation been rediscovered in the subject of gravitational collapse. The now notorious black holes have appeared with their own brand of time asymmetry.

Even quantum mechanics cannot escape our attentions. Ever since the work of von Neumann it has been realized that the act of measurement carried out on a quantum system is of a fundamentally different character than an ordinary interaction between two quantum systems. During the measurement there is an irreversible change which defines a time asymmetry. The history of attempts to associate this asymmetry with the thermodynamic properties of the measuring apparatus, or even the entire universe, is a long and complicated one.

Recently, a bizarre new twist has been given to our subject with the discovery that the elementary particle known as the neutral K meson violates time reversal symmetry. For the first time a physical system has been found which behaves asymmetrically in time as a consequence of an *interaction* rather than boundary conditions.

The intention of this book is to take the reader systematically through this entire range of topics. Clearly, it would be impossible to cover in detail every dissenting point of view in a single book. Instead, the emphasis here is on the presentation of a coherent, consistent overall picture. Extensive references are given at each stage to alternative theories and interpretations. Much of the subject matter has been developed in a cosmological context, because this has been the main area of interest in recent years.

Chapter 1 explains in greater detail the subject matter to be discussed, and develops carefully the appropriate terminology, definitions and philosophical platform. The standard work of Maxwell and Boltzmann's kinetic theory as far as the H theorem is traced in chapter 2, while chapter 3 is reserved for a full discussion of the reversibility objections and their resolution using the branch system concept of Reichenbach. Chapter 4 starts with a brief introduction at an elementary level to modern cosmology, Olbers' paradox and the expansion of the universe. The role of gravitation in thermodynamics is then discussed, along with the importance of the various elementary particle processes which occur in the big bang for understanding the present disequilibrium of the cosmos. Chapter 5 deals with the subject of wave motion, and retardation in electrodynamics. Much attention is devoted to the absorber theory, which is presented in some detail, including a determination of the various cosmological models which satisfy the 'absorber requirement'. Chapter 6 is devoted to quantum matters: a quantum treatment of statistical mechanics and a necessarily speculative account of the measurement theory. Tachyons and K mesons also get a brief mention. The final chapter, in more speculative vein, conjectures on the so-called heat death of the universe and the attempts by various cosmologists to avoid it. A section on some consequences of closed time is also included.

Having completed this book, it is felt by the author that a reasonably convincing account of all the many and varied aspects of time asymmetry can be consistently given, without running into the contradictions and paradoxes so often encountered in existing literature on the subject. This is only true if attention is restricted to the understanding of real phenomena, and excursions into essentially philosophical questions such as the 'subjectivity' of the asymmetry are avoided. There are, of course, still some gaps and loose ends in the account, and more seriously, no question of complete rigour in the mathematical sense. However, the basic framework for understanding the nature of time asymmetry is clearly available, and it is unlikely that any future considerations of detail will introduce anything qualitatively new.

Whether or not we can also meaningfully identify the *origin* of the asymmetry is a subtle point. However, it emerges from the following chapters that the event referred to as the *beginning* of the universe (in the big-bang sense) may be described as the temporal extremity at which there is a *mismatch* between the boundary conditions on the global dynamical motion as determined by gravity, and the microscopic particle motions of the cosmological material, as determined by electromagnetic, weak and strong interactions. In other words, there is no correlation between the large and small scale motions of the universe. In that case, it turns out that entropy will increase for the overwhelming majority of initial conditions of the cosmological material. Therefore this mismatch might be regarded as the ultimate origin of time asymmetry in the world, and for the ever expanding big-bang models, the mismatch simply accepted as a fact of nature, like the mass of the proton. In the case of cosmological models which either have no beginning, or which have *two* temporal extremities (a 'beginning' and an 'end') this acceptance is not so satisfactory. For example, consider the case of the closed Friedmann universe which expands from a singularity for half its life, then recontracts symmetrically to a final singularity. There is clearly then no reason why one extremity rather than the other should be singled out for the mismatched boundary conditions. For this reason, it is tempting to postulate that the microscopic boundary conditions are restricted to a very special subset which results in overall time symmetry between the singular points. Whether these remarks constitute an *explanation* or merely a *description* of time asymmetry, we leave to the philosophers.

Finally, a few words must be said about background reading. It is helpful if the reader has a working knowledge of special relativity and electrodynamics, though a brief account of the Lorentz transformation and proper time is given in chapter 1, and Maxwell–Lorentz theory reviewed in chapter 5. A knowledge of contravariant and covariant tensor labels is assumed, with indices being raised and lowered by the metric tensor, e.g. $x_\mu = g_{\mu\nu}x^\nu$. The usual convention that repeated indices are summed over is used throughout. Only Greek letters are used to label tensor components, Latin letters being reserved for particle labels, etc. Very little thermodynamics or statistical mechanics is assumed, the standard theory being presented where necessary. General relativity is kept to a minimum, much of the cosmological discussion being referred to local neighbourhoods of the comoving frame, in which ordinary language and definitions suffice. In chapter 6, a reasonable ability in quantum mechanics is demanded for the first section, though the discussion of quantum statistical mechanics and measurement theory requires only a rudimentary acquaintance with the subject.

Units will be used in which Planck's constant and the speed of light are both unity ($\hbar = c = 1$) unless stated to the contrary. In the electrodynamics Gaussian units are used. A list of symbols is given on pp. xi–xviii.

References are given at the end of each chapter. In addition, publications for further or background reading are listed. These include both review articles and books of general interest, as well as references to articles on specific topics connected with the material of the chapter, and not included in the references.

References

1. For example: 'Gain in entropy . . . is a subjective concept'. R. M. Lewis, *Science*, **71**, 569, 1930; 'Interpretations of irreversible equations of motion in classical statistical mechanics are "subjective".' H. D. Zeh, On the irreversibility of time and observation in quantum mechanics, in *Foundations of Quantum Mechanics* (Ed. B. D'Espagnat), Academic Press, New York and London, 1971.
2. A. S. Eddington, *The Nature of the Physical World*, Cambridge University Press, Cambridge, 1929, chapter 5.
3. J. Mehra and E. C. G. Sudarshan, *Nuovo Cimento*, **11 B**, 215, 1972.

4. L. D. Landau and E. M. Lifshitz, *Statistical Physics*, 2nd edition, Pergamon Press, Oxford, 1968, p. 31. See also M. Born, *Natural Philosophy of Cause and Chance*, Clarendon Press, Oxford, 1949, p. 113.
5. T. Gold, *Amer. J. Phys.*, **30,** 403, 1962.
6. F. Hoyle and J. V. Narlikar, *Proc. Roy. Soc.* A, **270,** 334, 1962. See also Time-symmetric electrodynamics and cosmology, in *The Nature of Time* (Ed. T. Gold), Cornell University Press, Ithaca, 1967.
7. J. M. Blatt, *Prog. Theor. Phys.*, **22,** 745, 1959.

1 Preliminary Concepts of Time

1.1 Time as a fourth dimension

The physical world of our experience is four dimensional. One of these dimensions is qualitatively distinct; it is called *time*. The other three dimensions are called *space*, while the whole four dimensional structure is referred to as *space–time*. Mathematically, space–time may be described by a differentiable *manifold* or continuum, for which every point, called an *event*, may be labelled by four independent quantities, or *coordinates*, written x^μ, where the index μ runs from 0 to 3.

The concept of *distance* or *interval* on the manifold may be expressed through its *metrical structure*. That is, the interval ds between two neighbouring points x^μ and $x^\mu + dx^\mu$ may be written in the form of a line element or *metric* as follows:

$$ds^2 = g_{\mu\nu}(x)\, dx^\mu\, dx^\nu. \tag{1.1}$$

The tensor $g_{\mu\nu}(x)$ is called the *metric tensor*, and is a function of the four position coordinates x^μ only. The type of geometry described by (1.1) is known as *Riemannian* geometry.

The precise form of $g_{\mu\nu}$, and hence the geometry of the manifold, is determined by the *gravitational* properties of the world in the neighbourhood of the point x^μ, in accordance with Einstein's general theory of relativity (see section 4.2). This geometry will normally be very complicated, but under some circumstances the metric reduces to a relatively simple expression. For example, in a small local region of space–time where gravity may be neglected, the manifold may be considered to be approximately *flat*. Under these conditions it is always possible to find a mapping such that the metric (1.1) assumes the simple form

$$ds^2 = +(dx^0)^2 - (dx^1)^2 - (dx^2)^2 - (dx^3)^2 \tag{1.2}$$

which is known as the *Minkowski* metric. It corresponds to the metric tensor

$$\eta_{\mu\nu} = \begin{bmatrix} +1 & & & \\ & -1 & & \\ & & -1 & \\ & & & -1 \end{bmatrix}. \tag{1.3}$$

Equations (1.2) and (1.3) show that one of the four coordinates is distinguished from the others by the positive sign. This special coordinate x^0 is the time, while $(x^1x^2x^3)$ refer to three dimensional space. The group of coordinate transformations that preserves this simple quadratic form of the metric is known as the *Poincaré* group. It comprises the translations and rotations, together with the so-called *Lorentz transformations* of the special theory of relativity, which connect the coordinates of reference frames which are in uniform relative motion. For example, consider two frames I and II in relative motion along the x^1 axis only, with uniform velocity v. Denote by $\bar{x}^\mu$ the coordinates in II; these are related to the x^μ in I by the following four equations:

$$\bar{x}^0 = \frac{x^0 - vx^1}{(1 - v^2)^{\frac{1}{2}}} \tag{1.4}$$

$$\bar{x}^1 = \frac{x^1 - vx^0}{(1 - v^2)^{\frac{1}{2}}} \tag{1.5}$$

$$\bar{x}^2 = x^2 \tag{1.6}$$

$$\bar{x}^3 = x^3. \tag{1.7}$$

It may be readily verified from equations (1.4)–(1.7) that the two intervals $(x^0)^2 - (x^1)^2 - (x^2)^2 - (x^3)^2$ and $(\bar{x}^0)^2 - (\bar{x}^1)^2 - (\bar{x}^2)^2 - (\bar{x}^3)^2$ are equal. Reference frames like I and II, which clearly form a privileged class, are often called *inertial* frames in special relativity.

It will be noticed that the right-hand side of (1.2) may be either positive, negative or zero; therefore, it is referred to as an *indefinite* metric. Two points which are separated by a positive, negative or zero ds^2 are referred to as being at *timelike*, *spacelike* or *null* separation respectively. (The terminology stems from the fact that a reference frame can then always be found for which only the time or space components of the interval survive, respectively.) In this way the manifold surrounding every point O divides naturally into three distinct regions (see figure 1.1): region $\mathscr{I} + \mathscr{I}'$, defined as the set of all points joined to O by a timelike interval; region $\mathscr{J}$, defined as the set of all points joined to O by a spacelike interval; finally, region $\mathscr{L} + \mathscr{L}'$ which are points at zero ds from O and known, for obvious reasons, as the *null cone*. It is

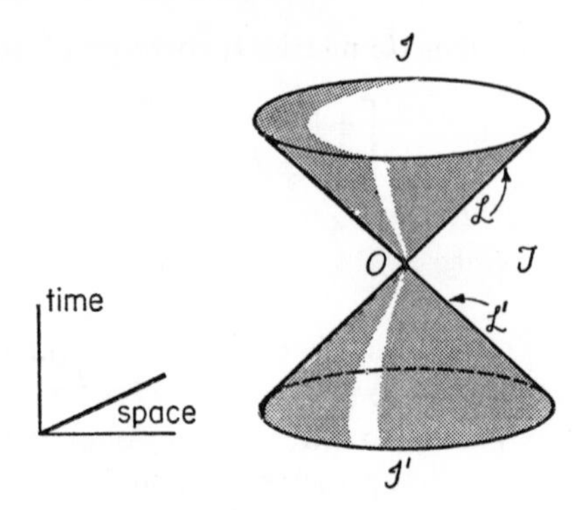

Figure 1.1 The regions of the manifold around O. One spatial dimension has been suppressed.

immediately obvious from figure 1.1 that the topology of the region $\mathscr{I} + \mathscr{I}'$ is fundamentally different from that of $\mathscr{J}$. The former consists of two disjoint sets of points, $\mathscr{I}$ and $\mathscr{I}'$, which may only be connected with a line which somewhere crosses the null cone.

The significance of this division of space–time at a point lies in the fact that it is preserved under Lorentz transformations. This is because the null cone is mapped onto itself, and the regions $\mathscr{I} + \mathscr{L}$ and $\mathscr{I}' + \mathscr{L}'$ are separately mapped onto themselves, so that the two halves of the null cone and their interiors are physically distinct. For this reason they are called the *absolute future* and *absolute past* of O, respectively. Of course, it is not yet supposed that there is any *structural* difference between these two halves. To call one the 'past' and the other 'future' is merely a convention corresponding to assigning increasing numbers to the x^0 axis up the page in figure 1.1. This procedure no more endows a physical asymmetry to the time axis than arbitrarily marking one of the space axes with increasing numbers; the significance of the former lies in its absolute character.

An event which is in the region $\mathscr{I} + \mathscr{L}$ of O we should say is *later* than O, because this event will always be found to be in the future of O in another reference frame; that is, all observers would agree on the earlier–later relationship of points with a timelike separation. On the other hand, an event in the region $\mathscr{J}$ which is at a later time than O for one frame, may well be at an *earlier* time in another (see figure 1.4). Thus the earlier–later relationship of events with a spacelike separation is meaningless.

We shall often have occasion to use the important notion of a *spacelike hypersurface* (or simply 'surface'), defined as a three dimensional surface in which all points are separated from all others by a spacelike interval (see figure 1.2). A line drawn everywhere orthogonal to a family of spacelike hypersurfaces is called a *timelike line*. It is found that all massive particles move along timelike lines in space-time; these lines are known as the *world lines* of the particles. (The possibility of particles with spacelike world lines, called tachyons, will be discussed in section 6.5.)

The world line of a typical particle in Minkowski space has been drawn in figure 1.2. The line has been marked with a sequence of points 1, 2, 3 ..., which may be referred to as a *chronological sequence* in view of the timelike character of the line. Each point has a definite past–future relationship with the others which is preserved under Lorentz transformations, so the whole sequence is the same for all inertial observers. This may be conveniently expressed by attaching a timelike vector tangentially to the world line. If the vector is moved along the world line either way it will always point in the same time direction, so that it may be used to define a consistent past–future relation along the entire line. By convention the direction in which the vector points may be called the 'future'. Repeating this procedure with other world lines enables a past–future convention to be established throughout the whole of space–time.

Although the remarks of the last paragraph were only applied to inertial observers, it is clear on continuity grounds that they may be extended to non-inertial observers also.

1.2 Newtonian time

Referring to figures 1.1 and 1.2, it will be seen that the null cone has been drawn at 45° to the time axis. This reflects the fact that all four dimensions of

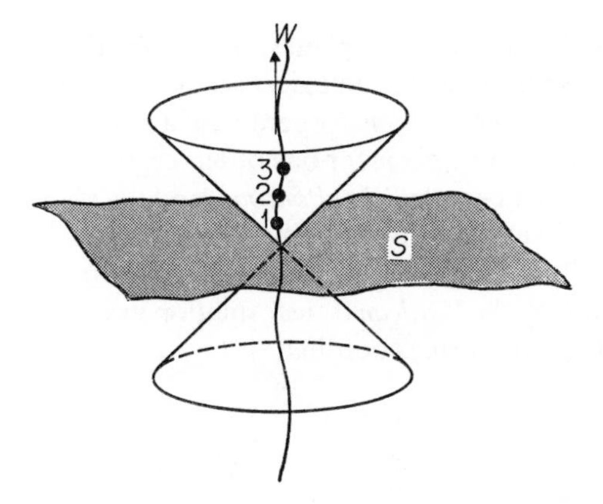

Figure 1.2 A spacelike surface S and a timelike world line W, marked with an ordered sequence of points. All observers agree on the ordering.

the manifold are being measured in the same units. Velocities are then dimensionless quantities. Now everyday velocities are rather small numbers and it is often convenient to introduce a large scale factor (the speed of light) in the time measurement. This has the effect of opening out the null cone to a very wide angle. Under these circumstances it is often a good approximation to consider the two halves of the null cone actually coinciding in a double plane sheet perpendicular to the time axis.

If we envisage a collection of such planes on top of one another, the planes may be labelled in a chronological sequence according to the world line points that they intersect (see figure 1.3). Because all observers see the same null cone

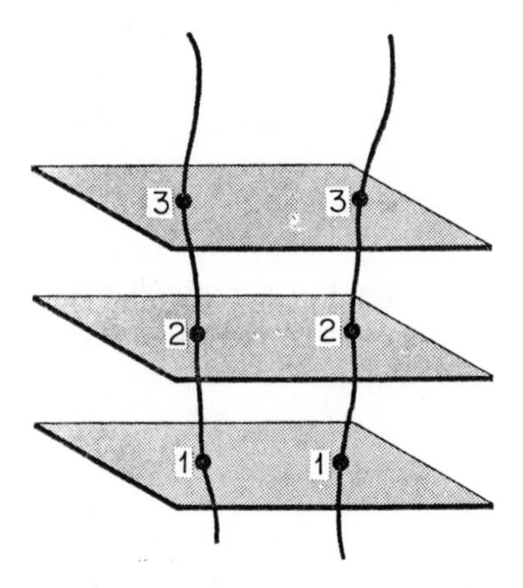

Figure 1.3 Planes of simultaneity formed by opening out the null cone. This was Newton's view of time.

they will all see the same labelled planes, which may therefore be used to label other world lines also in a chronological sequence. These surfaces then serve the function of *planes of simultaneity* because if a number of points on different world lines lie in the same plane for one observer, they do so for all observers. The points may then be called *simultaneous*, and the planes used to coordinatize an *absolute* invariant universal time, common to all reference frames.

Mathematically, this approximation corresponds to replacing the Lorentz transformation L by the *Galilean* transformation in which the time coordinate is uncoupled from the space coordinates:

$$\bar{x}^0 = x^0 \tag{1.8}$$
$$\bar{x}^1 = x^1 - vx^0 \quad \text{etc.}$$

Under these circumstances the old Newtonian theory of space and time is recovered. The null cones in relativity theory are defined physically by the geodesics of massless particles (e.g. photons). In pre-relativity days it was expected that the velocity of these particles would be frame dependent in accordance with the Galilean transformation (1.8), but in fact their velocity is the same in all frames because of the invariance of the null cone on which they move.

1.3 Clock rates and proper time

Although it was mentioned in section 1.1 that the chronological ordering along a world line is preserved under Lorentz transformations, nothing was said about the *spacing* of the sequence of events as determined from different inertial frames. To discuss this matter, it is convenient to draw the space–time coordinates of two reference frames, called I and II, on the same diagram.

In figure 1.4 the usual rectangular reference frame for system I has been drawn, with only one spatial axis, x^1, shown for simplicity. The x^0 axis may be thought of as the world line of a particle at rest at the origin of spatial coordinates in frame I. Frame II is assumed to be in motion along the x^1 direction with uniform velocity v, so that the time axis in this frame, labelled $\bar{x}^0$, is the world line of a particle moving with uniform velocity v in I. Now we come to the space coordinates. An observer in frame I regards the x^1 axis as

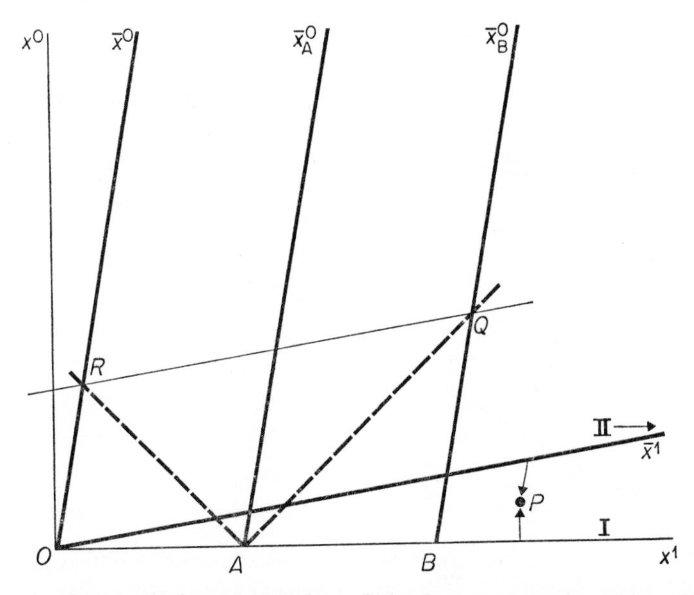

Figure 1.4 Geometrical representation of the Lorentz transformation. The null cone AR, AQ is common to both frames. Note that event P, which is spacelike with respect to O, is later than O in frame I but earlier than O in frame II.

his line of simultaneity. To find on the diagram (figure 1.4) the corresponding line of simultaneity for an observer in frame II, the following argument may be used. Consider two other equidistant world lines which are parallel to $\bar{x}^0$; they are called $\bar{x}^0_A$ and $\bar{x}^0_B$ in figure 1.4. The null cone at the point A on the x^1 axis is common to both reference frames because it is Lorentz invariant. It intersects the world lines $\bar{x}^0$ and $\bar{x}^0_B$ at points R and Q respectively. Now from the point of view of an observer in frame II the world lines $\bar{x}^0$ and $\bar{x}^0_B$ represent particles at rest an equal spatial distance from $\bar{x}^0_A$; hence the events R and Q are regarded as simultaneous. Thus the $\bar{x}^1$ axis is drawn parallel to this line of simultaneity through O.

We are now in a position to compare the time spacing of two events as seen from frames I and II. First it may be noted that devices exist which have the property of marking off systematically points at equal intervals *as measured along the world lines*. These devices are called *clocks*. The ticks of a clock define an ordered sequence along the world line. If a standard clock, i.e. one which ticks reliably at a preassigned rate, is carried in each frame, then a comparison of the interval between two ticks, and hence a comparison of the clock *rates*, may be made from observations in each frame.

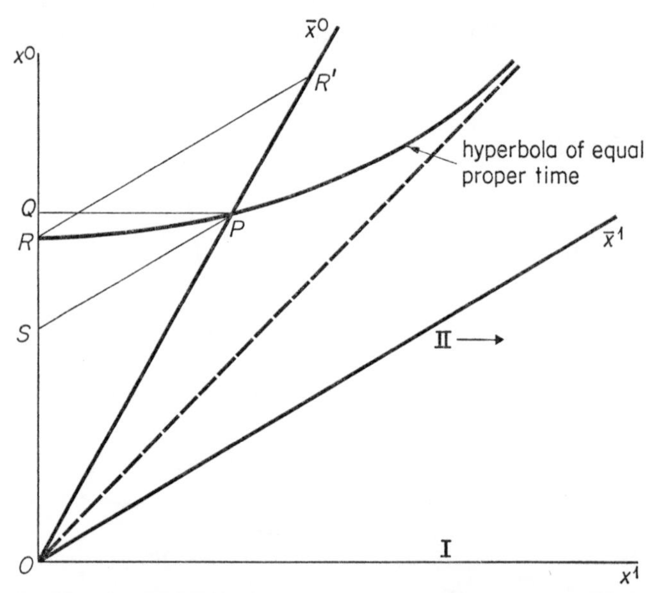

Figure 1.5 The time dilation effect. Standard clocks carried in frames I and II tick at events R and P respectively. In frame II the tick at R is reckoned to be simultaneous with R', i.e. after the tick at P, while in frame I the tick at P is reckoned to be simultaneous with Q, i.e. after the tick at R. Each observer sees his colleague's clock running slow.

To understand this comparison qualitatively, refer to figure 1.5. Before any results are interpreted, it must first be appreciated that although both reference frames have been drawn once again on the same diagram, the *scale* is different in either case. It will be recalled that we are working with an indefinite metric, so that distances measured along a null cone are zero, however far we go. Now the null cone through O bisects both pairs of axes, but because those of II are more oblique and so closer to the null cone, the scale for them is larger.

To determine the scale, note that the interval s is common to both frames. Choosing the unit interval $s^2 = 1$, we obtain from the form of the Minkowski metric the relations $(x^0)^2 - (x^1)^2 = (\bar{x}^0)^2 - (\bar{x}^1)^2 = 1$. This is the equation of a hyperbola with the null cone as asymptote, and is also drawn in figure 1.5. A standard clock which would mark off the unit interval OR in frame I would therefore mark off the interval OP in frame II.

Allowing for the obliqueness of the $\bar{x}^1$ axis, it will be seen that an observer in II reckons the end of this unit interval of time to be simultaneous with event S on the x^0 world line, the line PS being a line of simultaneity for this frame. On the other hand, an observer in I reckons the event P to occur simultaneously with event Q, because QP, which is parallel to the x^1 axis, is a line of simultaneity in *this* frame. In this way the observers will obviously disagree on the duration of the tick, and hence of the clock rates. The observer in I sees the standard clock in II running slow, because his own standard clock ticks at R, whereas II's clock ticks simultaneously with the *later* event Q. But likewise, the observer in II sees the clock in I running slow also, for his own standard clock ticks at P, whereas I's clock ticks simultaneously with the *later* event R'. This effect, which is clearly perfectly reciprocal between the two frames, is known as *time dilation*. Note that the delay in the ticks is not simply due to the delay caused by the finite travel time for the light by which each observer sees the other's clock (which, of course, can be allowed for).

The actual amount of dilation is readily found from the Lorentz transformation (1.4). An interval Δx^0 in I appears to be dilated to $\Delta \bar{x}^0$ in II, where

$$\Delta \bar{x}^0 = \frac{\Delta x^0}{(1 - v^2)^{\frac{1}{2}}}. \tag{1.9}$$

It is apparent from the above discussion that the notion of intervals of time is highly ambiguous. However, it is possible to define a unique time for any given reference frame, whatever the state of motion. This is the time as registered by a clock carried with the frame, and is referred to as the *proper time*. Only when we happen to be at rest with respect to this frame does the proper time coincide with the coordinate time x^0.

1.4 Cosmic time

It has been seen how the metric (1.1) can always be reduced in small regions of space–time to the Minkowski form (1.2). On the large scale, we do not expect the manifold to be flat because gravity can no longer be neglected. Over distances of, say, a galaxy, it is expected that the manifold geometry will be very complicated. However, it is generally accepted by astronomers that on the very large scale, i.e. over sizes of the order of clusters of galaxies, the gross features of the universe are *uniform*. This fact enables a metric to be constructed for the large-scale aspects of the universe which is also of a relatively simple form.

In constructing such a metric, account must be taken of the important fact that the universe appears to be *expanding*, so that the world lines of the galaxies fan outwards. It was first suggested by Weyl[1] that the smoothed out distribution of matter in the universe could be described by a bundle (or congruence) of non-intersecting timelike world lines, diverging from a common point in the past. It is then possible to find a one parameter family of spacelike hypersurfaces which are everywhere orthogonal to the world lines. A coordinate system may be chosen in which $\frac{dx^1}{ds} = \frac{dx^2}{ds} = \frac{dx^3}{ds} = 0$. This means that the matter is everywhere at rest relative to this coordinate system, which is therefore described as *comoving*. The family of surfaces may be labelled by the proper time of the world lines that thread through them, and because the surfaces are orthogonal to *all* the world lines these labels serve as a type of universal or *cosmic time*. Such surfaces clearly possess some of the properties of the Newtonian planes of simultaneity.

The existence of a cosmic time is made more physically intelligible by appealing to the *cosmological principle*. This principle asserts that the large scale features of the universe are the same in all directions and at all points, i.e. the universe is homogeneous and isotropic. It follows that each point of space–time has a privileged velocity, at which an observer will see isotropic expansion. This is the comoving frame; an observer at rest in this frame can therefore label time by the sequence of states that the universe passes through on the large scale, as it expands. This time can obviously be made common to all points on the manifold because all comoving observers see the same sequence of states. It should be mentioned that the Earth itself appears to be very nearly comoving.

From these symmetry properties alone it can be shown that the metric (1.1) may be written in the simple form

$$ds^2 = dt^2 - \frac{R^2(t)}{(1 + \frac{1}{4}kr^2)^2}(dr^2 + r^2\, d\Omega^2) \tag{1.10}$$

in radial polar coordinates (r, Ω). The right-hand side of (1.10) is known as the *Robertson–Walker* metric, after its discoverers[2]. The coordinate t is the cosmic time, and R is a scalar function of t only. The three-space surfaces of constant t have the metric

$$ds^2 = -\frac{R^2(t)}{(1 + \frac{1}{4}kr^2)^2}(dr^2 + r^2\,d\Omega^2) \tag{1.11}$$

so that $R(t)$ merely acts as a scale factor for the distance between any two comoving points. If $R(t)$ is chosen to be an increasing function of t, ds^2 increases with time, and the Robertson–Walker metric describes a universe which is in a state of *uniform expansion*. The precise form of $R(t)$ must be determined by the field equations of general relativity (section 4.2).

The geometry of the surfaces t = constant is determined by the factor k, which is only permitted to take the three values $+1$, -1 and 0. If $k = +1$, the curvature of the three-space is positive (spherical space). The space is then closed, and has finite volume, analogous to the surface of a sphere of two dimensions. In such a space, the surface area of a sphere of radius r is less than $4\pi r^2$. If $k = -1$, the curvature is negative, and the volume infinite (hyperbolic space), the surface area of a sphere being greater than $4\pi r^2$. Finally, $k = 0$ corresponds to a flat (Euclidean) space with the 'usual' geometric property that the surface area of a sphere is equal to $4\pi r^2$.

1.5 The physical meaning of time asymmetry

In section 1.1 it was mentioned that if a vector is attached tangentially to a world line, the direction of the vector is the same for all observers. This vector may then be used to define the absolute *future direction* as, say, the direction of the head of the arrow in figure 1.2. It is most important to appreciate that this assignation in no way implies an anisotropy or asymmetry of the space–time manifold, or the world line (or anything else) with respect to time. There is nothing *physical* about the distinction between past and future directions at this stage—they are merely convenient labels used for discussing different orientations in space–time unambiguously.

In sharp contradistinction to the remarks of the previous paragraph, in addition, it may also be considered that there exists a *physical* timelike asymmetry in space–time. This might be one of two varieties. (1) An asymmetry of the manifold itself with respect to time. For example, there is indeed a pervasive global asymmetry in virtue of the *expansion* of the universe, so that the scale factor R in the Robertson–Walker metric (1.10) defines this asymmetry through the sign of dR/dt. (2) An asymmetry of the world lines in virtue of their *collective* local behaviour. For example, diverging wavefronts of light or temperature equalizing heat flows may be preferentially oriented in time, either over the whole manifold, or in subregions of it. Some authors have even suggested that (1) and (2) might be strongly connected[3].

If we want to represent these asymmetries on a diagram, it might be convenient to do so by drawing an *arrow* in a timelike direction. Which direction does not matter; for instance, the head of the arrow could be drawn to point in the direction of outgoing spherical light waves. According to the choice, the arrow which represents any particular asymmetry will be either parallel or antiparallel to the other, *different*, arrow which was previously used to define the earlier–later or past–future direction convention. Thus, if the sensible rule is made that having fixed the earlier–later arrow, it remains the same for the whole manifold (as supposed in section 1.1) then it may be that at some particular point the arrows are parallel. With the conventions which have been selected above, this will actually be true in our own region of space–time in the real world. Nevertheless, it may certainly be conjectured that this parallelism will not be true at all places in the universe, or at all times. In general there will be several arrows to denote different physical asymmetries that seem to occur independently in nature, and we must allow for the possibility that all these arrows, or perhaps only some of them, may reverse with respect to the earlier–later arrow in some regions of the manifold. For example, although dR/dt defines a future-pointing arrow (expanding universe) at this time, this arrow may be reversed, i.e. past point, eventually (recontracting universe).

Having distinguished the real, physical asymmetry with respect to time from the *language* used to discuss temporal orientation on the manifold, it is important to also guard against muddling the former with the controversial phenomenon of *psychological time*. Conscious awareness of time is a *two-fold* experience. Firstly, the mind analyses time into a past and future *direction*, i.e. renders experience asymmetric with respect to time. For example we *remember* the past, but only *predict* the future. This *mental* lopsidedness is readily comprehensible in terms of the *physical* asymmetry of our environment when it is appreciated that our brains participate in the usual physical processes of the surrounding world, and are thereby constrained by the same asymmetries as any other electrochemical system. In short, our experience is asymmetric because we are coupled to other asymmetries in our environment. Specifically, the accumulation of information, called *memory* when it resides in our brains, is but a typical example of the general accumulation of information which is taking place all around us in the universe (e.g. craters on the moon give information about its past history). This is entirely in accordance with the informational interpretation of entropy in thermodynamics (see section 2.6).

As well as memory, there is a *second* aspect to our experience of time. Not only do our minds distinguish between the past–future *directions* of time from the physical asymmetry (and would presumably *reverse* with this asymmetry if such occurred), they also *divide* time into *the* past and *the* future. These two psychologically distinct parts of time are separated by the transient *present moment* or *now* which, it is often claimed, *moves* steadily from the past" towards the 'future'. It is this that gives rise to the impression that time 'passes'. For many people this impression is so profound that it ceases to be an aspect of mental activity and becomes instead a property of *time itself.* References are frequently made to the 'flux' or 'flow' *of* time, often exploiting the analogy of a forward flowing river[4].

The four dimensional space–time of physics makes no provision whatever for either a 'present moment' or a 'movement' of time. Indeed, since the concept of simultaneity has been explicitly overthrown by the special theory of relativity, there cannot in any case be any meaning to the 'same moment' in two different places. A universal now cannot exist in Minkowski space, where extension in space implies extension in time also. Rather than thinking in terms of a succession of experiences by a particular particle, we must instead deal with its entire world line in four dimensions; in the words of H. Weyl[5] 'the objective world simply *is*, it does not *happen*'.

In spite of this, many writers have asserted the objective physical existence of the now[6]. Sometimes the indeterminacy of quantum mechanics is invoked for supportive evidence[7]. Perhaps the most persuasive argument against the use of the concepts 'present moment' and 'flow of time' in physics (as opposed to psychology) is that they say nothing new; they are totally redundant concepts within the current framework of space–time. (Atoms possess *collective* qualities which are irrelevant to their individual properties. For example, the study of physiology does not lead us to suppose that the atoms of our bodies are alive. In the same way a study of psychology does not imply the participation of the atoms of our brains in the 'now' phenomenon. These remarks are valid notwithstanding the fact that we cannot 'draw a line' between the animate and inanimate[8].) In addition, they actually appear to be meaningless in this context, as has been emphasized by many authors. Indeed, it is hard to see how any statement concerning the movement of time, or of the now *in* time, can be other than a mere tautology in Minkowski space[9]. Just as it is meaningless to discuss the motion of a particle in three dimensional space without introducing the additional dimension of time (for example, compare the observation of a passing meteor with an examination of a still photograph of its track), so it is presumably meaningless to discuss motion in time, or of time, without introducing an additional dimension, or 'hyper-time'[10]. It may be considered excessive deference to human psychology to erect especially a five dimensional structure, though the reader may well be struck by the resemblance of such a conjecture to the Wigner interpretation of quantum mechanics (see section 6.3). In this interpretation the human mind also stands 'outside' the ordinary four dimensional universe (as though in an extra dimension), and in which the time asymmetry of the quantum measurement process is a consequence of the 'entering into the human consciousness' of the outcome of the measurement[11]. It is, of course, precisely this entry into the consciousness of the surrounding world with which we identify the transient 'present'.

Much of the controversy about time asymmetry has arisen over the confusion between the two quite distinct concepts of the real physical asymmetry of the world with respect to time, and the apparently *illusory* forward flow of psychological time[12]. This confusion has often led to the arrow, which was used previously to represent the *asymmetry* of some physical process with respect to time, being misinterpreted as the *direction of movement of the now*, i.e. the direction of the discredited flow of time. Whence comes the entirely misleading though universal practice of referring to time asymmetry as 'the arrow' of time, or even 'the direction' of time[13].

In addition to all this, there is a widespread belief that the origin of time asymmetry must be sought in the *structure of time itself*, that is, that intervals of time possess an *intrinsic* preferred orientation. Sometimes distinctions are even made between 'one-way' and 'two-way' time[14]. These speculations about oriented time are conspicuous in their inability to divulge anything new or useful about time, and are totally without empirical support. They are presumably a result of too close an identification with psychological time.

In sharp contrast to this *active* view of time, where time itself is endowed with questionable properties such as asymmetry and forward flow, the *passive* view will be adopted. It will be assumed that time has *no* intrinsic orientation, movement, direction or arrow whatever. Time for us is simply a subspace of the space–time manifold, the use of which enables events to be coordinatized chronologically. Indeed, time asymmetry exists in the real world. This will always be taken to mean the *asymmetry of the world with respect to time*, often abbreviated to 'asymmetry in time' or simply 'time asymmetry'. The last phrase must not be construed to mean an asymmetry *of* time.

1.6 The nature of the structural asymmetry. Time reversal invariance

It is a conspicuous fact of nature that the real world exhibits a structural difference between the two directions of time. That is, certain physical processes occur which are apparently asymmetric between these directions. The meaning of this statement will now be investigated a little more closely.

Time as a physical parameter was first introduced in a mathematical sense by Galileo, and later incorporated at a fundamental level into the laws of

Newtonian mechanics. It is clear from Newton's definition of time that he viewed the concept in terms of an absolute 'ever-flowing stream', something closely related to our physiological and psychological notion of the 'passage' of time. Although in section 1.5 a sharp distinction was drawn between the *flow* of time and *asymmetry* in time, for Newton there was no such distinction. Now when Newton formulated his laws of mechanics, it was found that the time parameter t (from now on, t rather than x^0 will usually be written) often entered into the equations symmetrically in the following sense.

Consider a point particle of mass m, described by a position vector $\boldsymbol{r}$, moving in a field of force $\boldsymbol{f}$. Furthermore, suppose that $\boldsymbol{f}$ does not explicitly depend on t (conservative field of force) or* $\dot{\boldsymbol{r}}$ (no velocity dependent forces), so that $\boldsymbol{f} = \boldsymbol{f}(\boldsymbol{r})$. Then Newton's second law of motion states

$$m\ddot{\boldsymbol{r}} = \boldsymbol{f}(\boldsymbol{r}). \tag{1.12}$$

Equation (1.12) may be solved subject to certain boundary conditions. For example, we may specify the position $\boldsymbol{r}$ and velocity $\dot{\boldsymbol{r}}$ at some initial time t_0, in which case the solution of (1.12) will give the trajectory of the particle for all time t. Suppose that one such solution for a particular set of boundary conditions is denoted by $\boldsymbol{r} = \gamma(t)$; then γ is a solution of (1.12):

$$\ddot{\gamma}(t) = \frac{1}{m}\boldsymbol{f}[\gamma(t)]. \tag{1.13}$$

It may be said that $\boldsymbol{r} = \gamma(t)$ is a dynamically possible trajectory for the particle in the field $\boldsymbol{f}(\boldsymbol{r})$. Now, if t is *formally* replaced by $-t$ in equation (1.13) we obtain

$$\ddot{\gamma}(-t) = \frac{1}{m}\boldsymbol{f}[\gamma(-t)]$$

$\left(\text{remembering } \frac{d^2}{dt^2} = \frac{d^2}{d(-t)^2}\right)$. Therefore $\gamma(-t)$ is a dynamically possible trajectory if $\gamma(t)$ is. This result is expressed by saying that, under these restricted circumstances, Newton's second law is *invariant under time reversal:* $t \to -t$. (The references to 'time reversal' are purely mathematical statements, and have nothing to do with a return to the past. It is to be identified physically with process or velocity reversal.)

We must be careful to realize that although $\gamma(-t)$ satisfies the equations of motion, it does not in general satisfy the boundary conditions, for $\gamma(-t_0) \neq \gamma(t_0)$ except in the trivial case of $\dot{\boldsymbol{r}} = 0$ for all t. The action of time reversal leaves $\boldsymbol{r}$ unchanged but reverses all velocities:

$$\boldsymbol{r} \to \boldsymbol{r}$$
$$\dot{\boldsymbol{r}} \to -\dot{\boldsymbol{r}}.$$

* A dot denotes total differentiation with respect to t.

$\gamma(-t)$ is therefore a solution of (1.12) with the *time reversed* boundary conditions of $\gamma(t)$. Time reversal may be imagined as taking a movie film of the original motion, and then playing it backwards. For example, a particle moving in a clockwise orbit becomes a particle moving in an anticlockwise orbit. If this time reversed motion is also a permissible solution of the equations of motion, then the system is said to exhibit *time reversal symmetry* or *reversibility*. The discussion can be immediately generalized to many particle systems. (Reversibility is always present when the Lagrangian of the system is invariant under time reversal.)

In elementary mechanics there are many examples of systems that do *not* possess the property of time reversal symmetry. A well-known example of this is the motion of a charged particle with charge e in a magnetic field, **B**. Such a system is described by the Lorentz law of motion

$$m\ddot{r} = e\dot{r} \times \mathbf{B} \tag{1.14}$$

which changes under time reversal to

$$m\ddot{\mathbf{r}} = -e\dot{r} \times \mathbf{B}. \tag{1.15}$$

To see what this means physically look at figure 1.6. A charged particle

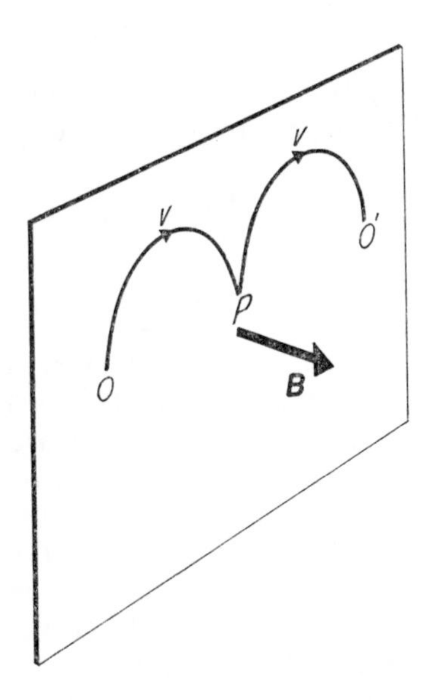

Figure 1.6 Violation of time reversal symmetry? The positively charged particle is projected from O at speed v, and travels along the arc to P under the force due to the magnetic induction $\boldsymbol{B}$. A reversal of the particle velocity at P will cause it to travel along the new path PO', and *not* backwards along PO.

initially at O is projected in the plane perpendicular to **B** with a speed v in some direction. It is well known that this particle describes the arc of a circle, and eventually reaches a point P in the same plane with the same speed v. If, at point P, the particle has its velocity reversed, it will *not* travel backwards along the curve PO, but will instead describe a different arc PO', as shown. A movie film of motion along PO would therefore conflict with the laws of electrodynamics. The system does not possess time reversal symmetry.

Another familiar example concerns the motion of a body through a viscous medium, or across a rough surface which exerts friction. For such systems it is usually found that the viscous drag, or frictional force, is proportional to $\dot{\mathbf{r}}$, the velocity of the body. The equation of motion (1.12) must then be replaced by

$$m\ddot{\mathbf{r}} = \mathbf{f}(\mathbf{r}) - \alpha\dot{\mathbf{r}}, \qquad \alpha = \text{positive constant.} \tag{1.16}$$

Equation (1.16) may be solved for the usual *damped* motion characterized by the situation $\mathbf{f}(\mathbf{r}) = 0$ (no driving force). In this case equation (1.16) may be immediately integrated to give $|\dot{\mathbf{r}}| \propto e^{-(\alpha/m)t}$, which has the property that $|\dot{\mathbf{r}}| \to 0$ as $t \to \infty$; the particle is brought to rest by the damping. On changing t to $-t$, the sign of the damping term in equation (1.16) is changed and, instead, exponentially growing solutions $|\dot{\mathbf{r}}| \propto e^{(\alpha/m)t}$ are obtained, which are clearly unphysical as they correspond to a body being spontaneously accelerated to an infinite velocity as a result of their contact with a viscous or frictional medium.

A different, intriguing type of damping force, discussed at great length in chapter 5, concerns the motion of a charged particle coupled to its *own* electromagnetic field. In that case the appropriate equation of motion can be shown to be, in the non-relativistic limit

$$m\ddot{\mathbf{r}} = \mathbf{f}(\mathbf{r}) - \tfrac{2}{3}e^2\dddot{\mathbf{r}}. \tag{1.17}$$

In equation (1.17) the third term represents the damping, but depends on the *third* derivative of $\mathbf{r}$ with respect to the time. The loss in kinetic energy of the particle appears as retarded electromagnetic radiation flowing away into space. The opposite situation, in which the $\dddot{\mathbf{r}}$ damping term is positive, corresponds to the convergence of advanced radiation onto the charged particle, and does not seem to occur in nature.

These three examples of asymmetric mechanical behaviour might be thought to provide some fundamental laws which indicate a selection of time directions. It is a remarkable fact that this is not so.

To understand this, the precise meaning of 'the system' in the above examples must be investigated. In all three cases, we only considered the mechanics of a single body moving in interaction with various agencies such as electromagnetic fields and viscous media. No inquiry was made about the detailed processes which produce the interactions. The time reversal discussed was only applied to the single degree of freedom of the body of interest and not to the whole system. If, instead a time reversal is applied to the entire system, the asymmetry encountered in the above three examples disappears.

In the first example the behaviour of a charged particle in a magnetic field was not reversible. However, the magnetic field itself is produced by moving charges, and under time reversal of these charges the field changes sign, $\mathbf{B} \rightarrow -\mathbf{B}$. If this transformation is included in equation (1.15) the time symmetry is restored.

In the second example of viscous or frictional damping, the motion of the body is slowed by the communication of kinetic energy to the medium atoms in the form of *heat*. It follows that if the motions of the individual atoms are also reversed then, because of the invariance of the laws of physics governing the atomic interactions, each collision will be reversed, causing a cooperative transfer of momentum to the large body, which would then become exponentially accelerated.

In the final example of radiative damping by an accelerated charged particle, it is supposed (pending the discussion in chapter 5) that the electromagnetic radiation which flows away from the vicinity of the charged particle is subsequently absorbed in the walls of the laboratory. The absorption occurs because the fields set the charged particles of the walls into motion. These moving charges then collide (reversibly) with other charges, and the energy becomes dissipated as heat. A time reversal of the whole system must now include reversing the motion of all the atoms of the laboratory walls, which will eventually give up their energy to charged particles. These accelerated charges will radiate, causing a coherent converging electromagnetic wave to collapse onto the original charged particle (which is accelerated thereby) in the time reversed fashion of the usual behaviour.

These three examples indicate what is, with one exception, a remarkable fundamental fact of nature: *all known laws of physics are invariant under time reversal.* In the three examples given, the reversibility is a consequence of the invariance of the laws of electrodynamics under time reversal (because even the atomic collisions are basically electromagnetic in nature). However, the same invariance is also found to apply to strong and weak interactions, and to gravitation, and this invariance is not removed by passing to a relativistic or quantum mechanical treatment.

This remarkable symmetry appeared to remain unbroken until recently, in 1968, when it was discovered that it was violated by certain elementary particle processes which involve the K meson. These processes do not appear to be at all relevant to the symmetry properties of most phenomena of interest in this book, so the subject of K mesons will be set aside until section 6.4, when a brief discussion will be given.

Although we are forced to conclude that the laws of physics do not themselves provide a time asymmetry, it is one of the most fundamental aspects of our experience that, as a *matter of fact*, the world is asymmetric in time. This is sometimes expressed by saying that the temporal asymmetry is 'fact like' rather than 'law like', or 'extrinsic' rather than 'intrinsic'. This means that asymmetric behaviour is observed as a result of the natural selection of certain types of special *boundary conditions* in preference to others. For example, although the possibility exists within the laws of physics for the exponential acceleration of a body by 'negative friction', the boundary conditions for this process do not *in fact* seem to occur in nature. Why this should be so will be explored in the chapters which follow.

References

1. H. Weyl, *Phys. Z.*, **24,** 230, 1923; *Phil. Mag.* (7), **9,** 936, 1930.
2. H. P. Robertson, *Astrophys. J.*, **82,** 284, 1935; **83,** 187, 257, 1936. A. G. Walker, *Proc. Lon. Math. Soc.* (2), **42,** 90, 1936.
3. T. Gold, *Amer. J. Phys.*, **30,** 403, 1962.
4. Thus we have a typical description due to J. Wild: 'Everything in the world seems to be engulfed in an irreversible flux of time which cannot be quickened or retarded, but flows everywhere at a constant rate'. *Rev. Metaphys.*, **7,** 543, 1954.
5. H. Weyl, *Philosophy of Mathematics and Natural Science*, Princeton University Press, Princeton, 1949, p. 116.
6. H. Reichenbach, *Ber. Bayer, Akad. Munchen. Math.-Naturwiss., Abt.*, **157,** 1925; *Philosophy of Space and Time*, Dover, New York, 1958, p. 138. M. Capek, *The Philosophical Impact of Contemporary Physics*, New York, 1961, p. 165. A. S. Eddington, *The Nature of the Physical World*, Cambridge University Press, Cambridge, 1929, chapter 5.
7. H. Reichenbach, *Ann. Inst. Poincaré*, **13,** 154, 1953. H. Bondi, *Nature*, **169,** 660, 1952. G. J. Whitrow, *The Natural Philosophy of Time*, Nelson, London, 1961, p. 295.
8. Contrast the remarks about dogs, cockroaches, etc, in *The Nature of Time* (Ed. T. Gold), Cornell University Press, Ithaca, 1967, p. 152.
9. A. Grünbaum, The anisotropy of time, in *The Nature of Time* (Ed. T. Gold), Cornell University Press, Ithaca, 1967, p. 152.
10. See for example R. Taylor, Spatial and temporal analogies and the concept of identity, in *J. Philos.*, **52,** 1955, reprinted in *Problems of Space and Time* (Ed. J. J. C. Smart), Macmillan, New York, 1964, p. 388. Also, how *fast* does time move? See J. J. C. Smart, *Analysis*, **14,** 80, 1954. M. Black, *Analysis*, **19,** 54, 1959.

11. A. Grünbaum, The anisotropy of time, in *The Nature of Time* (Ed. T. Gold), Cornell University Press, Ithaca, 1967, p. 152.
12. For example P. W. Bridgman, *Reflections of a Physicist*, Philosophical Library, New York, 1950, p. 163.
13. This point has been emphasized by J. J. C. Smart, *Analysis*, **14,** 80, 1954, and M. Black, *Analysis*, **19,** 54, 1959.
14. S. Hwang, *Foundations of Physics*, **2,** 315, 1972.

Further reading

1. A historical discussion of the concept of cosmic time is given in J. D. North, *The Measure of the Universe*, Clarendon Press, Oxford, 1965.
2. The literature concerning the status of the 'now' is immense. The following publications tend to support the point of view taken in this book, and contain many other references: A. Grünbaum, The status of temporal becoming, *Ann. N.Y. Acad. Sci.*, **138,** 374, 1967; *Modern Science and Zeno's Paradoxes*, Wesleyan University Press, Middletown, 1967, chapter 1; *Philosophical Problems in Space and Time*, Knopf, New York, 1963 and Routledge & Kegan Paul, London, 1964. D. Williams, The myth of passage, *J. Philos.*, **48,** 1951. J. J. C. Smart, The river of time, reprinted in *Essays in Conceptual Analysis* (Ed. A. G. N. Flew), Macmillan, London, 1956. See also the articles by Grünbaum, R. Taylor and N. Goodman in *Problems of Space and Time* (Ed. J. J. C. Smart), Macmillan, New York, 1964, together with the bibliography.
3. Some of the basic conceptual and terminological confusions of our subject are clarified in an article by M. Bunge, Time asymmetry, time reversal, and irreversibility, *Studium Generale*, **23,** 562, 1970.

2 Thermodynamics and Statistical Mechanics

2.1 The second law of thermodynamics

The subject of thermodynamics lies at the foundation of any discussion of processes which are asymmetric in time. There are two reasons for this. Firstly, because asymmetric phenomena of a thermodynamic character pervade the whole of our experience; indeed, our very existence as biological organisms is one example. Secondly, because the development of statistical mechanics, and the more recent theories of irreversible non-equilibrium processes[1], has opened the way to a clear qualitative understanding of the nature of the underlying asymmetry, even though rigorous mathematical proofs of much of it are still lacking.

The appearance of temporal asymmetry in thermodynamic phenomena is extraordinarily varied. For example, suppose that we remove the stopper from a flask filled with coloured gas, and watch it diffuse away; or place an ice cube in a jar of warm water and observe it melting; or witness the radiation of heat by an electric wire. In all these processes there is an asymmetry in the following sense: the reverse sequence of events is never seen to occur. The spontaneous accumulation of gas atoms in an opened flask is not encountered, nor the spontaneous freezing of a small part of warm water in a jug, nor the generation of electricity by heating up an electric wire.

Examples of this sort are unlimited, and range across our whole experience in every field of physics. Nevertheless, although chosen from such different classes of phenomena, all these examples have one thing in common. They display an asymmetry in time by destroying the usefulness of energy, in particular, heat energy. A great deal of attention is paid these days to the subject of energy resources available to man. But the existence of energy as such is of no use to us. Energy in *disequilibrium* is required before it can be put to good purpose. A frequently quoted example is man's inability to run a ship on the heat energy of the ocean. The sea water is at an absolute temperature of several hundred degrees, and yet man is incapable of removing this colossal energy for his own purposes. In all asymmetric thermodynamic phenomena we are witnessing the transition from energy in disequilibrium to useless equilibrium, from which no change can occur.

These experiences are summarized in a statement known as the second law of thermodynamics, which has been stated in many forms. The form due to Lord Kelvin and Clausius says, more or less, that heat does not, of its own accord, flow from cold to hot bodies. The qualification 'of its own accord' is crucial here, for refrigerators by the degradation of electrical or chemical energy indeed bring about such a flow. In short then, all systems left to themselves (isolated) tend to approach thermal equilibrium and not to leave it again. As this is clearly a 'one way in time' tendency (i.e. moving nearer to equilibrium as we pass from past to future, in our usual convention), it supplies an asymmetry to the space–time manifold as discussed in chapter 1. Later we shall examine the question of whether this asymmetry is in the same direction across the whole space–time manifold, or whether there may be some regions where it is absent or even reversed. For the moment it will be assumed that a consistent direction to the asymmetry in our local region of space–time is a well-established fact of life, and we will proceed to inquire how this situation might come about.

The rather heuristic remarks of the previous paragraphs may be made more precise by the introduction of the first and second laws of thermodynamics. The identification of the heat, Q, with *energy* enables it to be related to the *internal energy*, E, of a thermodynamic system, and the *work done*, W, by that system on its surroundings in any change. This relation is embodied in the *first law of thermodynamics:*

$$dE = đQ - đW. \tag{2.1}$$

For the moment the discussion will be restricted to the important special case of a system enclosed in a volume V, possessing uniform pressure p, temperature T and composition throughout. E is a function only of the state of the system, in this case say p and V. The change in E when passing from one state to another does not therefore depend on the path taken, but only on the initial and final states. Hence dE in equation (2.1) is a perfect differential. This is not so for $đQ$ and $đW$, which are imperfect differentials (denoted by a horizontal bar through the d), i.e. their change in value between two states does depend on the path taken, so that they may not be considered to be the differentials of some functions Q and W (see figure 2.1). However, $đQ$ may be made into a perfect differential by multiplication with an appropriate integrating factor. Historically, this integrating factor was inferred from certain considerations about the performance of heat engines. Specifically, if a path is followed which has the form of a closed loop, then there are certain

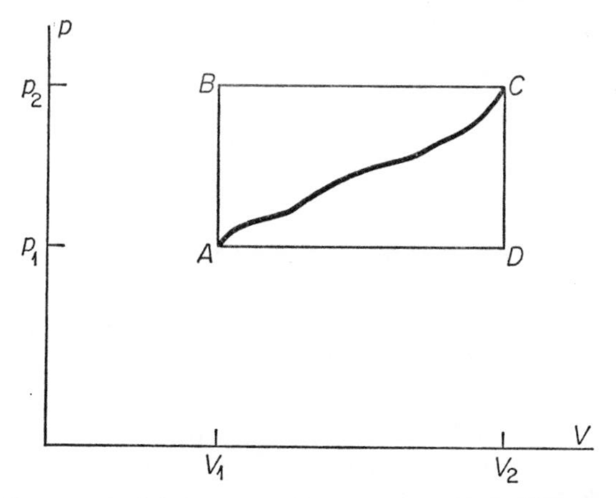

Figure 2.1 Three paths joining the two states (p_1, V_1) and (p_2, V_2) of a gas. Along the upper path (ABC) the pressure is raised from p_1 to p_2 while the volume remains unchanged (no work done), after which the gas is expanded from volume V_1 to V_2 at constant pressure p_2, with an amount of work $p_2(V_2 - V_1)$ done on the surroundings. Along the lower path (ADC) the gas is expanded first, so that the work done is only $p_1(V_2 - V_1)$. The middle path (AC) shows an intermediate case.

processes for which the integral $\oint \frac{dQ}{T}$ taken around the loop vanishes. Roughly speaking, these processes are such that the same path may be followed around in the opposite direction in such a way that both the system *and* its surroundings are restored to their original condition. That is, at every step the changes in the entire system and environment are *reversible*, a term that will be more carefully explained in due course. An obvious corollary of the above mathematical property is that the integral $\int \frac{dQ}{T}$ taken between two *different* points on the closed loop is the same whichever way around the loop one goes. As the shapes of the closed paths are entirely arbitrary, the latter is independent of the path chosen between the two points. Therefore, in the case of reversible processes, the quantity $\frac{dQ}{T}$ is a perfect differential, and the integrating factor for dQ may be identified with the reciprocal of the temperature.

The perfect differential so obtained is denoted by dS, where S is a new function of state called the *entropy*. Using dS, the first law of thermodynamics, equation (2.1), may be written

$$dS = \frac{dE}{T} + p\,dV$$

which also uses the fact that the work done in the simple system under consideration is just that due to mechanical expansion, $p\,dV$. In a more complex system in which we allow for changes in the internal relative concentrations of the different components, $n_1, n_2, n_3 \ldots$, this equation is replaced by

$$dS = \frac{dE}{T} + p\,dV + \sum_i \frac{\partial S}{\partial n_i}\,dn_i. \tag{2.2}$$

In the real world strictly reversible changes do not occur, though they may be approached as idealizations. All the examples of time asymmetric changes given at the beginning of this section were in contrast *irreversible*. That is, it would not be possible to return the system *and* surroundings to their original state. That does not imply that, for example, a melted ice cube could not be refrozen, but that in order to do so the necessary apparatus (such as a refrigerator) would necessarily undergo its own irreversible changes in order to operate. It is never possible to get *everything* back as it was without using something outside the system.

The general behaviour of irreversible processes may be described by the *second law of thermodynamics*, which may be stated in the form

$$\Delta S \geqslant \int \frac{đQ}{T}$$

where ΔS is the gain in entropy, and the equality applies only to strictly reversible processes. In the special case where the system is isolated inside an adiabatic (heat-impermeable) enclosure, $đQ = 0$, and the inequality reduces to

$$\Delta S > 0 \tag{2.3}$$

for any natural change which occurs in the real world. Expressed in words, the second law requires that *the entropy of an isolated system never decreases.* Furthermore, because all natural changes increase the entropy of an isolated system, a condition from which there is no further change must be one of *maximum entropy*. Such a condition is known as thermodynamic *equilibrium.*

To illustrate the second law, two examples will be considered.

Suppose thermal contact were established between two heat reservoirs 1 and 2 which are maintained at constant temperatures T_1 and T_2 respectively. If $T_1 > T_2$ it is expected that a quantity of heat, say Q, will flow from 1 to 2 because 1 is hotter. The entropy change for reservoir 1 is $-Q/T_1$ while that for T_2 is $+Q/T_2$. The total entropy change is therefore $Q\left(\frac{1}{T_2} - \frac{1}{T_1}\right)$, which in view of the difference in temperature is a quantity greater than zero. The entropy has therefore increased. In this example the second law of thermodynamics expresses the principle that heat always flows spontaneously from hot to cold bodies. Only if $T_1 = T_2$ (thermal equilibrium) is the heat flow reversible, because neither reservoir is hotter than the other. In that case the entropy change is zero as expected.

As a second example, consider N atoms of a monatomic ideal gas (no intermolecular interactions) enclosed inside a cylinder and piston which are adiabatic. It is known from kinetic theory that the entropy of an ideal monatomic gas is, to within an arbitrary additive constant,

$$S = Nk \log(VT^{\frac{3}{2}}), \tag{2.4}$$

k being Boltzmann's constant. There are two extreme ways in which the gas may be expanded. The first consists of withdrawing the piston extremely rapidly (much faster than the mean molecular speed) so that the gas essentially expands into a vacuum and does no work on the piston in the process. The temperature of the gas cannot change, because it is simply proportional to the internal energy for an ideal gas, and this will be constant by the first law. It then follows from equation (2.4) that the entropy of the gas will increase by an amount $Nk \log(V_2/V_1)$, V_1 and V_2 being the initial and final volumes respectively.

On the other hand, if the piston is withdrawn infinitesimally slowly, it may be assumed that at each instant the pressure and temperature remain uniform throughout the gas (equilibrium), and are related by the ideal gas law $pV = NkT$. In this case, the gas will do work on the retreating piston, of an amount given by the integral of $p\,dV = NkT\,dV/V$. Being an adiabatic enclosure, it may be deduced from the first law (equation (2.1)) that

$$dE = \tfrac{3}{2}Nk\,dT = \frac{NkT\,dV}{V}$$

from which is obtained the result that $\log(VT^{\frac{3}{2}})$ is a constant. Equation (2.4) then shows that the entropy change is zero. This was as expected because the latter method of expansion is *quasi*-static, and clearly reversible. If the piston were to be advanced again in the same quasi-*static* manner, the gas could be returned to its original state with the piston in the same position and no external work done or heat exchanged. This would not be so in the rapid expansion case, because in order that the process be reversed, the gas would need to anticipate the advance of the piston, and retreat before it so that the piston did no work against the gas pressure. In this second example, the second law becomes an expression of the principle that a gas will explode into a vacuum, but will never spontaneously implode into a smaller volume. Some insight is also obtained into the nature of the concept of reversibility. In order to approach the ideal reversible change, it is necessary for the constraints on a system to change on a time scale much longer than the characteristic *relaxation* time of the system (see section 2.2), for whatever process is initiated by the changing constraints. In the examples given it is clear that the reversible changes were those that took place under *equilibrium* conditions.

The use of the words 'reversible' and 'irreversible' can be confusing because, as will be shown in chapter 3, the results of statistical mechanics do indeed admit the possibility that the entropy of an isolated system decreases, when the individual molecular motions are taken into account. Nevertheless, when making statements about the *macroscopic* properties of thermodynamic systems in this book, these words will continue to be used. The reader should be cautioned not to confuse this *macroscopic* reversibility with the completely different topic of *microscopic* reversibility, or microreversibility, which will be discussed in section 2.2, and which is closely related to the subject of time reversal invariance or symmetry of the laws of physics (see section 1.6). Where ambiguity may arise, the adjectives 'microscopic' and 'macroscopic' will be used explicitly. The distinction is important because microscopic irreversibility does not necessarily imply an asymmetry in time, nor does microscopic reversibility imply symmetry in time. On the other hand, thermodynamic macroscopic irreversibility does imply an asymmetry in time.

2.2 The kinetic theory of gases

The first step towards understanding the nature of the temporal asymmetry involved in the second law of thermodynamics comes with the application of the laws of classical mechanics to the individual molecules from which a macroscopic system is composed. The pioneering work on this subject was carried out by Clausius[2], Maxwell[3] and Boltzmann[4]; an outline of the important developments will be considered here.

A simplified model of a gas will be taken as the system for consideration. It consists of a large number, N, of identical spherical particles confined to a volume, V, by perfectly smooth rigid walls, from which the particles bounce elastically. No energy may penetrate the walls, so that the inside of the box is an isolated, adiabatic enclosure. The particles may also collide with each other and interact via short range forces, but the gas is assumed to be sufficiently dilute that only binary collisions need be considered. The motion of the particles will be treated according to the laws of classical non-relativistic mechanics. The use of more sophisticated models, including quantum mechanical considerations (see chapter 6) does not introduce essentially new features into the subject of temporal asymmetry.

The complete microscopic configuration of our system is given by specifying the three position coordinates q_i, and three momentum coordinates p_i, of all N particles. Once these are given, the complete behaviour of the gas for all past and future times is determined through the causal equations of motion (initial value problem—see also section 5.7). These six coordinates may be used to construct a six dimensional space, known as μ space. The position and momentum (or velocity) of a particle at any time is associated with a point in μ space. The entire system of N particles is associated with N points in μ space. As the microscopic configuration of the gas changes, the N points will move about. The motion of individual points is not being considered because typically we are dealing with 10^{23} particles (one gram molecule). In addition, (1) the positions and momenta of all the particles simultaneously cannot be known, (2) even if they were, a calculation of their motions would be impossibly complex. Instead we proceed as follows. Divide up μ space into small cells so that the volume of the cells is large enough to contain a great many particles, but still small enough to be considered as infinitesimal compared to macroscopic dimensions. The size of the cells corresponds to the limits of resolution of macroscopic observation. For real gases the above two requirements are quite compatible.

Each cell is labelled by the average values of the coordinates, denoted by $\boldsymbol{q}, \boldsymbol{p}$ for short, and has a volume $\prod_{i=1}^{3} dq_i\, dp_i$, denoted by $d^3q\, d^3p$. From a macroscopic standpoint we are only concerned with the total number of particles in the cell, without distinguishing between them. This number is denoted by $f(\boldsymbol{q}, \boldsymbol{p}, t)\, d^3q\, d^3p$, where f is the *density* of points in μ space; it is called the *distribution function*. Integrating f over all the cells of μ space gives

$$N = \iint f(\boldsymbol{q}, \boldsymbol{p}, t)\, d^3q\, d^3p.$$

Knowing f, any macroscopic quantity of interest may be calculated by suitably averaging over the whole distribution.

To find the equation of motion for f, first suppose that the interaction between the particles has been switched off. In real space, each particle will then move in a straight line with constant $\boldsymbol{p}^2$. In μ space the points will move on the surfaces of hyperspheres; the cells being hyperspherical shells in this case. Under these circumstances, the N points will move independently on continuous trajectories inside these shells. These trajectories in μ space therefore bear a similarity to the streamlines of fluid in ordinary space; in particular, there will be a conservation equation, because N is constant

$$\frac{\partial f}{\partial t} + \boldsymbol{\nabla}_6 \cdot (\boldsymbol{u} f) = 0. \tag{2.5}$$

This equation is the usual conservation equation generalized to six dimensions: $\boldsymbol{u}$ is the six dimensional 'velocity' vector $(\dot{q}_1, \ldots \dot{p}_3)$ and $\boldsymbol{\nabla}_6$ is the six dimensional divergence $\left(\frac{\partial}{\partial q_1} \ldots \frac{\partial}{\partial p_3}\right)$. In the case under consideration $\boldsymbol{p}^2$ is a constant, and equation (2.5) reduces to

$$\frac{\partial f}{\partial t} + (\boldsymbol{v} \cdot \boldsymbol{\nabla}_q) f = 0 \tag{2.6}$$

with $\boldsymbol{v} = \boldsymbol{p}/m$ and ∇_q being the three dimensional velocity, and gradient operator (acting on $\boldsymbol{q}$) respectively. Equation (2.6) is sometimes generalized by including the effects of an external force $\boldsymbol{F}$ acting on the system, e.g. gravity. In that case, in place of equation (2.6) the following should be obtained

$$\left(\frac{\partial}{\partial t} + \boldsymbol{v} \cdot \nabla_q + \frac{\boldsymbol{F}}{m} \cdot \nabla_v\right) f(\boldsymbol{q}, \boldsymbol{v}, t) = 0 \tag{2.7}$$

where ∇_v is the gradient operator acting on $\boldsymbol{v}$.

For simplicity it will be assumed that $\boldsymbol{F} = 0$, and also that the gas is uniformly distributed throughout V (i.e. no correlation assumed between positions and velocities), so that $f = f(\boldsymbol{v}, t)$ and (2.6) becomes simply

$$\frac{\partial f}{\partial t} = 0. \tag{2.8}$$

f is therefore a constant of the motion for *any* distribution of velocities or energies. Expressed differently, any function f is an equilibrium distribution.

This being so, the following problem will now be considered. Imagine that we are able to throw the N points into μ space at random, the only restriction being that the total energy of the system is fixed, and equal to E. Each throw will produce a different microscopic arrangement of points, but many throws will appear to give the *same* distribution function f, because macroscopically it is not possible to distinguish between rearrangements within the small cells. It is easy to calculate the most probable distribution function under these circumstances. We simply calculate the total number of permutations consistent with a given distribution, and maximize this subject to the two constraints of fixed N and E. The calculation is performed using the method of Lagrange multipliers; the result is well known:

$$f(\boldsymbol{v}) \propto e^{-\beta v^2}, \tag{2.9}$$

which is known as the *Maxwell distribution.*

So far the discussion has been restricted to a perfectly ideal gas. Account must now be taken of the collisions between particles which occur when their interaction is switched on. The equation of continuity (2.5) was based on the assumption of continuous trajectories in μ space. If a particle suffers a collision, its associated point will be displaced from the cell that it occupies, and will no longer travel along with its neighbour points as before. The trajectory will end abruptly and reappear somewhere else corresponding to the new position and velocity of the particle. (It is assumed that the gas is sufficiently dilute that the collisions may be regarded as sudden rare interruptions to free particle motion.) Similarly, other points will be displaced from other regions of μ space by collisions, and reappear in the region of interest. This non-conservation of points in a given region must be taken into account by adding a source term to the right-hand side of equation (2.5). This term will be called $\left(\frac{\partial f}{\partial t}\right)_{\text{coll}}$ and it will also appear in (2.6), (2.7) and (2.8). The general equation will be

$$\left(\frac{\partial}{\partial t} + \boldsymbol{v} \cdot \nabla_q + \frac{\boldsymbol{F}}{m} \cdot \nabla_v\right) f = \left(\frac{\partial f}{\partial t}\right)_{\text{coll}} \tag{2.10}$$

which is known as *Boltzmann's equation.*

Physically, the effect of the collisions, however small the interactions, can be pictured as follows. Suppose initially we have an arbitrary distribution function $f(\boldsymbol{v})$; because of the collision term on the right-hand side of equation (2.8) this is no longer necessarily an equilibrium distribution. After each collision time, the points in μ space will be shuffled abruptly to new positions owing to the effect of the collisions. It is at this stage that a crucial, and unproved, assumption of a statistical nature must be introduced. It will be assumed that

the points in μ space are reshuffled at random. (A)

If assumption A is correct then the reshuffled points will be more likely to end up in a high probability distribution than a low probability distribution. In particular, they are most likely to end up in the most probable distribution (2.9). Once there, it is unlikely that subsequent reshuffling will remove them. The Maxwell distribution is therefore regarded as the natural equilibrium distribution for a gas. For N of order 10^{23}, the probability of substantial deviations from (2.9), once attained, is exceedingly small[5]. The time required to reach equilibrium sufficiently closely is known as the *relaxation time*, and is closely related to the time between collisions of the molecules.

In spite of the fact that the foregoing discussion provides a plausible physical picture of how a gas will progress, under the influence of random collisions, from an arbitrary state to an equilibrium state, this has not been demonstrated mathematically. The first step in a mathematical treatment is to evaluate the collision term $\left(\frac{\partial f}{\partial t}\right)_{\text{coll}}$ for the model under consideration. This calculation is considerably simplified by the symmetric shape of the particles, and the need to treat only binary collisions. It is more convenient to work with velocities rather than momenta. In this case, the number of particles, called type 1, which have velocities between $\boldsymbol{v}_1$ and $\boldsymbol{v}_1 + d\boldsymbol{v}_1$, and are located in a small region d^3q of space, is $f_1(q, v_1, t)d^3qd^3v_1$, which will be abbreviated to $f_1d^3qd^3v_1$.

Consider one such type 1 particle; it will not necessarily remain in the element $d^3qd^3v_1$ for it is all the while bombarded by other particles which may knock it out by collision. For example, type 2 molecules, with velocities between $\boldsymbol{v}_2$ and $\boldsymbol{v}_2 + d\boldsymbol{v}_2$, constitute a flux $f_2 \, |\boldsymbol{v}_1 - \boldsymbol{v}_2| \, d^3v_2$, where $|\boldsymbol{v}_1 - \boldsymbol{v}_2|$ is the relative velocity between the particles. Collisions with these parameters will change the incident velocities of such particles from $\boldsymbol{v}_1$, $\boldsymbol{v}_2$ to $\boldsymbol{v}_1'$, $\boldsymbol{v}_2'$, and turn the angle between them through an amount Ω. If the cross section for collision of these type 1 and type 2 particles is called $\sigma(\Omega)$ there will be

$$f_2 \, |\boldsymbol{v}_1 - \boldsymbol{v}_2| \, \sigma(\Omega) \, d^3v_2 \, d\Omega \tag{2.11}$$

collisions per unit time with one type 1 particle. Therefore, the total number of type 1 particles disturbed by collisions per unit time is

$$f_1 \int d^3v_2 \int d\Omega \, \sigma(\Omega) \, |\boldsymbol{v}_1 - \boldsymbol{v}_2| \, f_2 \tag{2.12}$$

where an integration has been taken over all velocities $\boldsymbol{v}_2$ and angles of deflection Ω.

In order to proceed further, it is necessary to make an assumption, which we shall refer to as *molecular chaos*. It was introduced by Boltzmann under the name *Stosszahlansatz*, but was already implicit in the work of Clausius. It may be stated as follows:

the positions and velocities of the particles are uncorrelated before they collide. (B)

Assumption B is independent of A, and the relationship between them will become clear in due course. Physically, condition B means that the number of type 2 particles rushing in to collide with type 1 are the same as the number of type 2 particles anywhere else in the gas, i.e. the incident flux does not depend on the fact that a collision is *going* to occur. Mathematically, it means that f_1 and f_2 are the *same function.*

Collisions which scatter particles back into $d^3q \, d^3v_1$ are called *inverse* collisions. The expression analogous to (2.12) for these processes will be

$$f_1' \int d^3v_2' \int d\Omega \, \sigma'(\Omega) \, |\boldsymbol{v}_1' - \boldsymbol{v}_2'| \, f_2' \tag{2.13}$$

where f_1' and f_2' denote the same distribution function as above, but with arguments $\boldsymbol{v}_1'$ and $\boldsymbol{v}_2'$ respectively. σ' is the cross section for inverse collisions.

So far nothing has been said about the nature of the collision forces. The precise nature of these forces is not of interest, but they are required to be *invariant under time reversal.* As stated in section 1.6, it does appear that this requirement is realized in nature. An additional requirement is that they be short range forces. If the collisions of the type under discussion are denoted by $\{\boldsymbol{v}_1, \boldsymbol{v}_2 \rightarrow \boldsymbol{v}_1', \boldsymbol{v}_2'\}$ and the inverse collisions by $\{\boldsymbol{v}_1', \boldsymbol{v}_2' \rightarrow \boldsymbol{v}_1, \boldsymbol{v}_2\}$, then the *reverse* collision is understood to be $\{-\boldsymbol{v}_1', -\boldsymbol{v}_2' \rightarrow -\boldsymbol{v}_1, -\boldsymbol{v}_2\}$, obtained by reversing the direction of all the particles. Invariance of the interaction under time reversal means that we may equate the cross section of any collision with its reverse:

$$\sigma(\boldsymbol{v}_1, \boldsymbol{v}_2; \boldsymbol{v}_1', \boldsymbol{v}_2') = \sigma(-\boldsymbol{v}_1', -\boldsymbol{v}_2'; -\boldsymbol{v}_1, -\boldsymbol{v}_2) \tag{2.14}$$

a result known as the classical *principle of microreversibility.*

Equation (2.14) may be used to obtain information about the cross section σ' for *inverse* collisions $\{\boldsymbol{v}_1', \boldsymbol{v}_2' \rightarrow \boldsymbol{v}_1, \boldsymbol{v}_2\}$. This is on account of the fact that the particles in our model possess additional symmetries: invariance under rotations and reflections in a plane. The combination of all three symmetries is equivalent to symmetry under inversion[6], i.e. we may equate

$$\sigma(\boldsymbol{v}_1, \boldsymbol{v}_2; \boldsymbol{v}_1', \boldsymbol{v}_2') = \sigma'(\boldsymbol{v}_1', \boldsymbol{v}_2'; \boldsymbol{v}_1, \boldsymbol{v}_2). \tag{2.15}$$

This result is not valid in the case of less symmetric particles, which may not possess kinematically allowed inverse collisions (see figure 2.2); this situation arises in classical considerations of polyatomic molecules. In these circumstances it is necessary to consider closed cycles of collisions[7]. However, in quantum mechanics, the equality (2.15) always applies, provided spin is considered to be reversed under time reversal (see chapter 6).

Finally, in addition to these symmetries regarding σ, the following result is also obtained from the kinematics of binary collisions:

$$|\boldsymbol{v}_1 - \boldsymbol{v}_2| = |\boldsymbol{v}_1' - \boldsymbol{v}_2'|. \tag{2.16}$$

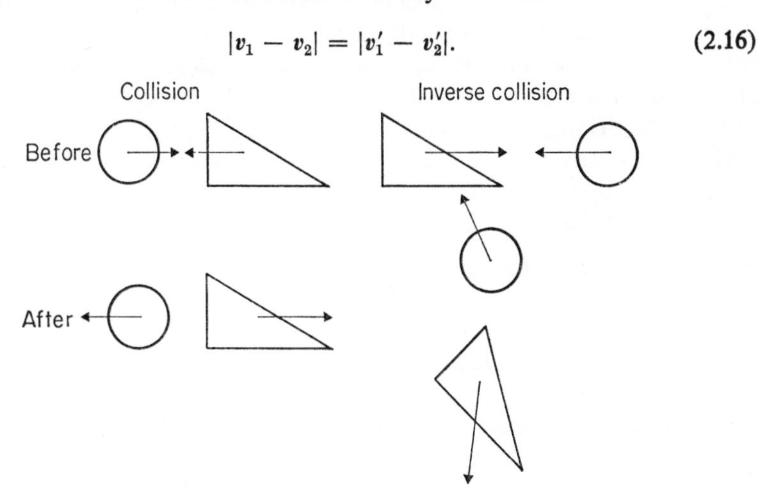

Figure 2.2 Classical collisions can violate reversibility. On the left is shown a collision between a sphere and a wedge, and on the right is the corresponding inverse collision, which is clearly very different. For spherical particles the additional symmetry ensures reversibility.

Moreover, it is easy to show

$$d^3v_1\, d^3v_2 = d^3v_1'\, d^3v_2'. \tag{2.17}$$

An integral expression for the collision term $\left(\frac{\partial f}{\partial t}\right)_{\text{coll}}$ may now be written down. To do this equation (2.12) is subtracted from (2.13) to obtain the *net* increase in points in the region $d^3q\, d^3v_1$ as a result of collisions, and then the three results (2.15), (2.16) and (2.17) used to simplify the expression obtained:

$$\left(\frac{\partial f_1}{\partial t}\right)_{\text{coll}} = \int d^3v_2 \int d\Omega\, \sigma(\Omega)\, |\boldsymbol{v}_1 - \boldsymbol{v}_2|\, (f_2' f_1' - f_2 f_1). \tag{2.18}$$

The Boltzmann equation (2.10) may then be written as

$$\left(\frac{\partial}{\partial t} + \boldsymbol{v}_1 \cdot \boldsymbol{\nabla}_q + \frac{\boldsymbol{F}}{m} \cdot \boldsymbol{\nabla}_{v_1}\right) f_1 = \int d^3v_2 \int d\Omega\, \sigma(\Omega)\, |\boldsymbol{v}_1 - \boldsymbol{v}_2|\, (f_2' f_1' - f_2 f_1). \tag{2.19}$$

2.3 The H theorem

In the last section two things were achieved. Firstly, a plausible physical argument based on assumption A was presented which could account for how a gas initially in an arbitrary condition might, under the influence of internal collision processes, progress steadily towards an equilibrium configuration. Secondly, the actual effect of collisions in the case of our simple model for a gas was explored, and the results expressed as an integrodifferential equation of motion for the distribution function, based on assumption B. It remains to show that (1) this equation really does result in f approaching the Maxwell equilibrium distribution (2.9) as $t \to \infty$, and (2) assumption B is valid.

Returning to the simplified case with $\boldsymbol{F} = 0$, and a spatially uniform density of particles throughout the volume V of the confining box, the condition for equilibrium, $\frac{\partial f_1}{\partial t} = 0$, may be written

$$\int d^3v_2 \int d\Omega\, \sigma(\Omega)\, |\boldsymbol{v}_1 - \boldsymbol{v}_2|\, (f_2' f_1' - f_2 f_1) = 0. \tag{2.20}$$

A necessary and sufficient condition that equation (2.20) be satisfied is

$$f_1 f_2 = f_1' f_2'. \tag{2.21}$$

Physically, the solution (2.21) expresses the fact that in equilibrium all types of collision processes are exactly balanced by their inverses. This is an example of the *principle of detailed balance*.

It follows from (2.21) that

$$\log f_1 + \log f_2 = \log f_1' + \log f_2' \tag{2.22}$$

which is to say that $\log f_1 + \log f_2$ is unchanged by collisions. Therefore, $\log f$ must be a linear combination of the kinematic quantities that are conserved in a binary collision, i.e. total momentum and energy. The most general form of (2.22) will then be

$$\log f = -\beta(\boldsymbol{v} - \boldsymbol{v}_0)^2 + \log C \tag{2.23}$$

where β, C and the components of $\boldsymbol{v}_0$ are arbitrary constants. In fact these constants determine some of the macroscopic properties of the gas; in particular $\boldsymbol{v}_0$ represents the velocity of the gas as a whole. For a vessel at rest $\boldsymbol{v}_0$ vanishes, so that (2.23) may be expressed in the form $f \propto \exp(-\beta \boldsymbol{v}^2)$, which is just the expected Maxwell distribution (2.9).

To show that f would tend towards this distribution from an arbitrary initial state, Boltzmann introduced his famous *H function*, defined here as

$$H = \int d^3v f(\boldsymbol{v}, t) \log f(\boldsymbol{v}, t). \tag{2.24}$$

It is easy to show that H is minimized by the Maxwell distribution (2.9); this is true independently of the assumption B of molecular chaos. H therefore measures the deviation of the system from equilibrium.

Differentiation of (2.24) gives

$$\frac{dH}{dt} = \int d^3v \frac{\partial f}{\partial t}(1 + \log f) \tag{2.25}$$

so we may substitute into (2.25) from the Boltzmann equation (2.19). Remembering that the second two terms on the left-hand side of (2.19) are zero in our model, equation (2.25) becomes

$$\frac{dH}{dt} = \int d^3v_1 \int d^3v_2 \int d\Omega\, \sigma(\Omega)\, |\boldsymbol{v}_1 - \boldsymbol{v}_2|\, (f_2' f_1' - f_2 f_1)(1 + \log f_1). \tag{2.26}$$

Now the integral in (2.26) is unchanged by interchange of the dummy labels 1 and 2, and interchange of the primed and unprimed variables; the latter follows as a consequence of (2.16) and (2.17). Taking the average of all four equal forms of (2.26) the following is obtained:

$$\frac{1}{4} \int d^3v_1 \int d^3v_2 \int d\Omega\, \sigma(\Omega)\, |\boldsymbol{v}_1 - \boldsymbol{v}_2| \times [f_2' f_1' - f_2 f_1] \times [\log(f_2 f_1) - \log(f_2' f_1')]. \tag{2.27}$$

The integrand of (2.27) is *never positive*; both terms in the square brackets change sign together. Indeed, there is the general inequality

$$(x - y)\log\frac{y}{x} \leqslant 0 \tag{2.28}$$

with equality only for $x = y$. In (2.27) this equality is just the condition for equilibrium (2.21).

We have therefore proved the important result

$$\frac{dH}{dt} \leqslant 0 \tag{2.29}$$

which is known as Boltzmann's H theorem. It states that when a gas is in a condition of molecular chaos, H will decrease; H will attain a minimum value at equilibrium (even without molecular chaos). H bears similar properties to the entropy S of thermodynamics. Both tend to change in one way only (S goes up, H goes down); both take extreme values at equilibrium (thermodynamic or statistical). Actually, if the constants β and C that occur in (2.23) are evaluated, one obtains

$$H = -\frac{N}{V}\log(VT^{\frac{3}{2}}) + \text{constant} \tag{2.30}$$

where the quantity T is proportional to the average energy per particle. If T is identified with the temperature of the gas, equation (2.30) may be interpreted as the usual definition of the entropy of an ideal gas, with

$$S = -kHV \tag{2.31}$$

where k is a constant of proportionality known as Boltzmann's constant. It has the value $1{\cdot}380 \times 10^{-16}$ erg K^{-1}.

2.4 Statistical mechanics

In the previous section Boltzmann's H theorem was outlined based on the assumption B of molecular chaos. If this assumption were always correct, this would amount to a proof that the H function for our model gas would decrease steadily from its initial value to its minimum value corresponding to equilibrium, as a result of molecular collisions. In real gases one would expect this equilibrium to be achieved fairly rapidly (fractions of a second, say). Such a demonstration would then amount to a 'derivation' of the second law of thermodynamics through the identification (2.31). The apparent recovery of a result which is manifestly asymmetric in time, from an analysis based on the principle of microreversibility, seemed paradoxical, and led to the so-called 'reversibility objections' to be considered in the next chapter. It is clear that the unrestricted H theorem must be wrong, and therefore the assumption of molecular chaos is called into question.

A powerful insight into the H theorem and its underlying statistical assumptions (at first sight rather obscure) is provided by the subject of statistical mechanics in the form due mainly to Gibbs[8].

Rather than work in μ space, a new space will now be considered, the coordinates of which are the $3N$ positions q_i and $3N$ momenta p_i of the N particles in our box. This $6N$ dimensional phase space of the *entire* system is known as Γ space. The reader should be most careful to distinguish between μ space and Γ space to avoid confusion. The complete microscopic condition of the gas is represented by a *single* point in Γ space. This representative point will move about as the system evolves; the motion is described by Hamilton's canonical equations

$$\dot{p}_i = -\frac{\partial \mathscr{H}}{\partial q_i} \qquad (2.32)$$

$$i = 1, \ldots 3N$$

$$\dot{q}_i = \frac{\partial \mathscr{H}}{\partial p_i} \qquad (2.33)$$

where $\mathscr{H}$ is the total Hamiltonian of the system

$$\mathscr{H} = \mathscr{H}(q_1, \ldots q_{3N}; p_1, \ldots p_{3N}).$$

It is well known that from the structure of the equations (2.32) and (2.33), an integration may in principle be performed, which will determine for all past and future times the values of all $6N$ positions and momenta, if they are completely specified at any one time. That is, if we know the location of the representative point in Γ space at any one time, the entire trajectory in Γ space is determined. Because the solution of (2.32) and (2.33) is *unique*, there will be just one possible trajectory through every point of the space. Such a trajectory must be either open, or a simple closed curve, for if the path intersects itself at a point, the representative point must move once again along the uniquely determined trajectory through that point. Because the system under consideration is isolated, and moves inside perfectly elastic walls, the total energy will be a constant of the motion, so that not all $6N$ coordinates in Γ space are independent. The representative point moves on a $6N - 1$ dimensional surface, called the energy surface. The volume of this surface is finite, because of the finite spatial extension of the system.

Fundamental to Gibb's approach is the introduction of what is called an *ensemble* of systems. In general, a macroscopic observation will be consistent with a very large number of different microscopic configurations, so that instead of considering just a single system, we now treat a very large number $\mathcal{N}$ of mental copies. Each copy will have an associated representative point moving in Γ space, so that the whole ensemble will be described by a cloud or swarm of points, all moving together in a complicated fashion. The ensemble members in no way *interact* with each other physically; they are a collection of mental rather than physical copies. Consequently, the points in Γ space move entirely independently, and should never be confused with the real physical motion of the N interacting molecules in ordinary space, or their associated points in μ space.

With this caution in mind, we now proceed to treat the swarm of points as a fluid in Γ space. To describe this fluid, a density ρ of points is introduced, analogous to the distribution function of μ space:

$$\rho = \rho(q_1, \dots q_{3N}; p_1, \dots p_{3N}, t) \tag{2.34}$$

ρ is in general a function of position in Γ space, and time t. The right-hand side of (2.34) will be abbreviated to $\rho(q, p, t)$. The total number of points $\mathcal{N}$ is given by taking an integral over the whole space

$$\mathcal{N} = \int \dots \int \rho(q, p, t)\, dq_1 \dots dp_{3N}. \tag{2.35}$$

$\mathcal{N}$ is assumed to be large enough so that ρ can be considered to change continuously to sufficient approximation.

The *average* value of any function $M(q, p)$ of the coordinates q and p (known as a phase function), taken over the ensemble at the time t, is

$$\bar{M}(q, p) = \mathcal{N}^{-1} \int \dots \int M(q, p)\rho(q, p, t)\, dq_1 \dots dp_{3N}. \tag{2.36}$$

(A single bar will always be used to denote ensemble averages.) Sometimes it is convenient to normalize ρ to unity, by replacing $\mathcal{N}$ by 1 in (2.36). ρ must then be regarded as a *probability* density for a single point picked out at random from the ensemble. In this case (2.36) must be replaced by

$$\bar{M}(q, p) = \int \dots \int M(q, p)\rho(q, p, t)\, dq_1 \dots dp_{3N}. \tag{2.37}$$

Since the number of points $\mathcal{N}$ is constant by assumption, the swarm may be considered as a fluid obeying a conservation equation, which may be written in a form similar to that of equation (2.5)

$$\frac{\partial \rho}{\partial t} + \nabla_{6N} \cdot (v\rho) = 0 \tag{2.38}$$

with $\boldsymbol{v}$ a $6N$ dimensional 'velocity' vector

$$\boldsymbol{v} \equiv (\dot{q}_1, \ldots \dot{q}_{3N}; \dot{p}_1, \ldots \dot{p}_{3N}),$$

and $\boldsymbol{\nabla}_{6N}$ a $6N$ dimensional divergence

$$\boldsymbol{\nabla}_{6N} \equiv \left(\frac{\partial}{\partial q_1}, \ldots \frac{\partial}{\partial q_{3N}}; \frac{\partial}{\partial p_1}, \ldots \frac{\partial}{\partial p_{3N}}\right)$$

in complete analogy with ordinary three dimensional space.

Expanding the second term of (2.38)

$$\boldsymbol{\nabla}_{6N}(\boldsymbol{v}\rho) = \sum_{i=1}^{3N}\left(\frac{\partial \rho}{\partial q_i}\dot{q}_i + \frac{\partial \rho}{\partial p_i}\dot{p}_i\right) + \sum_{i=1}^{3N}\rho\left(\frac{\partial \dot{q}_i}{\partial q_i} + \frac{\partial \dot{p}_i}{\partial p_i}\right). \tag{2.39}$$

The equations of motion (2.32) and (2.33) may be used to simplify the right-hand side of (2.39). The first term is clearly equal to

$$\sum_{i=1}^{3N}\left\{\frac{\partial \rho}{\partial q_i}\frac{\partial \mathscr{H}}{\partial p_i} - \frac{\partial \rho}{\partial p_i}\frac{\partial \mathscr{H}}{\partial q_i}\right\}$$

which is usually abbreviated as the Poisson bracket $\{\rho, \mathscr{H}\}$. The second term is seen to vanish on account of the relation

$$\sum_{i=1}^{3N}\left\{\frac{\partial \dot{q}_i}{\partial q_i} + \frac{\partial \dot{p}_i}{\partial p_i}\right\} = 0$$

which is a consequence of differentiating (2.32) and (2.33). Therefore, one obtains from (2.38) and (2.39)

$$\frac{d\rho}{dt} + \{\rho, \mathscr{H}\} = 0. \tag{2.40}$$

If the *total* derivative of ρ is considered as a function of the q, p and t, equation (2.40) reduces to

$$\frac{d\rho}{dt} = 0. \tag{2.41}$$

This fundamental result is known as Liouville's theorem, and is invariant under canonical transformations of the q, p. Physically, it implies that the density of points remains constant in the neighbourhood of any selected point which moves along with the fluid (the density does not in general remain constant at a fixed point in Γ space). The swarm of points moves as if it were an incompressible fluid.

There will be certain distributions of points which will have the special property that they are in *statistical equilibrium*. This occurs if

$$\frac{\partial \rho}{\partial t} = 0 \tag{2.42}$$

which is equivalent to the requirement that

$$\{\rho, \mathscr{H}\} = 0 \tag{2.43}$$

according to (2.40). Such a condition is satisfied if ρ is a function of any dynamical variable which itself is a constant of the motion.

A trivial case of a stationary ensemble is the *uniform ensemble*

$$\rho = \text{constant} \tag{2.44}$$

in which the points are distributed uniformly throughout Γ space. It follows immediately from (2.42) that this distribution remains constant in time. It is also invariant under canonical transformations.

There is another important example, which is applicable to isolated conservative systems, such as the closed box of gas discussed earlier. In this case the total energy $\mathscr{H}(q, p) = E$ is a constant of the motion, so that if ρ is chosen to be any function of E, this distribution will also be in equilibrium (stationary). (The reader will have noticed the close similarity between this discussion and that leading up to equation (2.9) for the function f in μ space. The formalism is in fact identical.) One such function is

$$\begin{aligned} \rho(E) &= \text{constant in the range } E \text{ to } E + \delta E \\ &= 0 \text{ outside this range.} \end{aligned} \tag{2.45}$$

The distribution (2.45) is known as the *microcanonical ensemble*. It is best pictured as a uniform density of points distributed across the energy surface only. If we consider there to be a natural uncertainty δE in a measurement of the energy E, then the surface is really a shell filled uniformly with points. The reason for the constancy of the microcanonical ensemble is clear if it is regarded as being formed from the uniform ensemble by discarding all the points outside the shell. As no point can cross a surface of constant energy anyway, this operation cannot affect the (equilibrium) distribution of points within the shell. Ensembles in statistical equilibrium are important for the representation of systems which are themselves also in equilibrium.

To understand the relevance of ensembles to the behaviour of a single physical system, first it must be pointed out that a measurement carried out on the system would necessarily take a finite interval of time T which would, moreover, be very long compared with the duration of microscopic processes. If it could be proved that the average behaviour of a system over an infinite period of time was the same as the equilibrium behaviour, it would follow that the system must be in equilibrium most of the time, and hence must return rapidly to equilibrium from any non-equilibrium state. (In section 2.2 the equilibrium state was regarded as the *most probable* condition for a gas; here it is treated as the *average* condition.) This would therefore be equivalent to proving the H theorem for the following reason. The H theorem predicts that a non-equilibrium state of a gas will soon pass to an equilibrium state which will persist, so that once again the average state of the gas over an infinite time is the equilibrium state.

This problem was also tackled by Boltzmann, who calculated the time average of a phase function $M(q, p)$, defined by

$$\hat{M} = \lim_{T \to \infty} \frac{1}{2T} \int_{t-T}^{t+T} M(q, p)\, dt \tag{2.46}$$

by dealing instead with the easier average taken over the microcanonical ensemble, i.e. $\bar{M}$ as defined by equation (2.36). The equality of the two averages $\hat{M}$ and $\bar{M}$ he took to rest on a hypothesis, known as the *ergodic hypothesis*, which states that:

the trajectory of a representative point passes eventually through every point on the energy surface. (C)

That the truth of hypothesis C does indeed lead to the equality of time and ensemble averages may be seen from the following considerations. It has already been mentioned that there is a unique trajectory through every point in Γ space. Therefore, if *one* trajectory passes through *every* point (as postulated by C), there can only be one possible trajectory. *Every* system, therefore, follows the same evolutionary path in Γ space, for wherever the representative point is located on the energy surface at a given time it will be on this unique trajectory, differing only by the time at which a given point is passed. Obviously then, the time average of a phase function taken over an infinite period is the same for all $\mathcal{N}$ systems in the ensemble; the actual trajectory need not be known to compute this average. Consequently, without change, the time average can be averaged over the ensemble, i.e. $\bar{M} = \hat{\bar{M}}$. But $\hat{\bar{M}}$ must be equal to $\bar{\hat{M}}$ (the time average of the ensemble average), because the order of averaging is immaterial. Finally, in view of the fact that the microcanonical ensemble is stationary, the time average in this latter quantity is superflous, and $\bar{\hat{M}}$ can be replaced by $\hat{M}$. This proves the assertion that $\bar{M} = \hat{M}$.

From the ergodic hypothesis it is possible to prove a slightly stronger property about the relative amounts of time that the representative point spends in different regions of the energy surface. The time spent in a given volume is in fact proportional to that volume, i.e. the point spends equal intervals in equal volumes.

If it were true, the ergodic hypothesis C would provide an exact dynamical justification for the use of representative ensembles in the description of single systems, without an additional statistical postulate. Unfortunately, it cannot be true in the form stated, because the points of a trajectory form a set of measure zero, which cannot therefore fill the energy surface whose measure is not zero. When it was realized that C was false, it was still hoped that most physical systems would satisfy the *quasi-ergodic hypothesis*, which asserts only that the representative point covers the energy surface everywhere densely, i.e. it passes arbitrarily close to any given point. Birkhoff[9] and von Neumann[10] have actually demonstrated that in a certain class of systems the time and ensemble averages are equal. Nevertheless, a general proof of this property for realistic physical systems is lacking. Moreover, it is easy to construct manifestly non-ergodic idealized models, such as a set of particles inside a smooth sided cubical box, moving on parallel paths perpendicular to one of the rigid faces. The particles would continue to bounce backwards and forwards for ever, thus confining the representative point to a small subset of the total energy surface. The time average for such a system is obviously very different from the ensemble average.

Attractive as the ergodic hypothesis might be, these difficulties appear to make the introduction of some tacit statistical assumption seem unavoidable (for example, we are not really interested in time averages over an infinite interval, as real measurements take only a finite time), so that some authors[11] prefer to let the validity of statistical mechanics rest on a bald statistical postulate, referred to as the fundamental postulate of statistical mechanics:

in the absence of any information about a given equilibrium system, the representative point is equally likely to be found in any equal volume of Γ *space.* (D)

Statement D, also known as the postulate of equal *a priori* probabilities, thus asserts that a sequence of measurements on a given, single system will tend to give the results predicted by the uniform ensemble average (see equation (2.44)). Nothing is asserted about the long time behaviour of this individual system, which may be very different from the average. In some cases we may have partial knowledge about the system; for example, in the case of our isolated box of gas we know (perhaps only approximately) the total energy. Postulate D then requires that we construct a representative ensemble with the points distributed as evenly as possible consistent with the partial knowledge of the system (this would be the microcanonical ensemble in the isolated case).

As regards the plausibility of assumption D, it is noted that the Hamiltonian equations do not themselves lead to a concentration of points in one part of Γ space or the other, so that some justification would be required for weighting different regions *unequally*. Furthermore, because of the invariance of the volumes of Γ space under canonical transformations, D is true in all canonical coordinate systems. Finally, a statistical assumption like D, although not derivable from the laws of mechanics, cannot contradict them.

If we wish to deal with a system which is not itself in equilibrium, it must be represented by a non-stationary ensemble, constructed in accordance with the fundamental assumption D. This leads the way to a *generalized H theorem* for ensembles that is free of the objections raised against Boltzmann's unrestricted *H* theorem for single systems.

2.5 The generalized $\overline{H}$ theorem

It was mentioned in section 2.4 that the stationary microcanonical ensemble can be used to calculate the values of phase functions $M(q, p)$ for a system in equilibrium. If it is known that the system is not in equilibrium, it may be described instead by an ensemble which changes with time, and approaches the microcanonical ensemble asymptotically. If a function σ in Γ space, analogous to the function H in μ space, is defined as follows

$$\sigma = \int \dots \int \rho \log \rho \, dq_1 \dots dp_{3N} \tag{2.47}$$

then it is easily proved that the microcanonical ensemble minimizes σ. On the other hand, if (2.47) is differentiated

$$\frac{d\sigma}{dt} = \int \dots \int \frac{d\rho}{dt} \{1 + \log \rho\} \, dq_1 \dots dp_{3N} \tag{2.48}$$

then the right-hand side is seen to vanish as a consequence of Liouville's theorem (2.41)

$$\frac{d\sigma}{dt} = 0. \tag{2.49}$$

σ does *not* therefore contain an asymmetric tendency to progress towards equilibrium.

A quantity which does have this property was introduced by P. and T. Ehrenfest[12], using the notion of *coarse graining*. It is not possible to know the exact location of the representative point in Γ space, because that would amount to a complete microscopic knowledge of the system. Our macroscopic knowledge is contained in the distribution function $f(\boldsymbol{q}, \boldsymbol{p}, t)$ in μ

space, which amounts to specifying the numbers of points n_i in each cell labelled i of μ space. To each set of numbers $\{n_i\}$ there will be a corresponding finite region Λ in Γ space, called a *star*. This is because the points in one cell of μ space may always be moved around inside the cell, and permuted, without changing the state of the system macroscopically. Of course, different macroscopic states will be associated with stars of different volumes. By assumption, the most precise location that a macroscopic observation may place on the representative point is that it resides within a given star. As the point moves about, it will wander from one star to another after a finite sojourn in each.

Having introduced the concept of stars, it is now possible to define a coarse-grained density P (capital rho) as follows

$$\mathrm{P} = [W(\Lambda)]^{-1} \int \dots \int_{\Lambda} \rho \, dq_1 \dots dp_{3N} \tag{2.50}$$

where $W(\Lambda)$ is the volume of the star Λ. ρ will be referred to as the fine-grained density, and σ as the fine-grained H function. P is just the mean of the fine-grained density over the star Λ; it is taken to be normalized to unity here. There is also a coarse-grained H function, denoted by $\bar{H}$, and defined by

$$\bar{H} = \int \dots \int \mathrm{P} \log \mathrm{P} \, dq_1 \dots dp_{3N}. \tag{2.51}$$

Since log P is constant over a given star, the right-hand side of (2.51) may be replaced by

$$\int \dots \int \rho \log \mathrm{P} \, dq_1 \dots dp_{3N} \tag{2.52}$$

so that $\bar{H}$ may be considered to be the average value of log P over the ensemble, according to equation (2.37). $\bar{H}$, like σ, is also minimized by the microcanonical ensemble.

Suppose that a macroscopic observation is made of a given isolated system at a time t_1 and an appropriate representative ensemble corresponding to this state is set up. In accordance with the postulate D of equal *a priori* probabilities, the fine-grained density ρ will be chosen to be constant over the regions of the energy surface consistent with this observation, and zero elsewhere. In this case, clearly $\mathrm{P} = \rho$, and $\bar{H}$ may be written at $t = t_1$

$$\bar{H}_1 = \int \dots \int \rho_1 \log \rho_1 \, dq_1 \dots dp_{3N}. \tag{2.53}$$

If equilibrium prevails at $t = t_1$, then the equality $\mathrm{P} = \rho$ will remain unchanged as the system evolves, so that $\bar{H}$ will not change either, in accordance with the result (2.49). If equilibrium does not prevail, then at a later time t_2,

P and ρ will no longer be equal, as ρ will have evolved to some new more uniform distribution (the reader should remember that at time t_1, ρ is only constant in the region of interest, and zero elsewhere—see also section 3.1). The value of $\bar{H}$ at this later time will be, according to expression (2.52)

$$\bar{H}_2 = \int \dots \int \rho_2 \log \mathrm{P}_2 \, dq_1 \dots dp_{3N}. \qquad (2.54)$$

The change in $\bar{H}$ between t_1 and t_2 is given by subtracting (2.54) from (2.53)

$$\bar{H}_1 - \bar{H}_2 = \int \dots \int \{\rho_1 \log \rho_1 - \rho_2 \log \mathrm{P}_2\} \, dq_1 \dots dp_{3N}. \qquad (2.55)$$

Now, $\rho_1 \log \rho_1$ may be replaced in the integrand of (2.55) by $\rho_2 \log \rho_2$, because of the equality (2.49). Moreover, the quantity $\mathrm{P}_2 - \rho_2$ will vanish when integrated over $dq_1 \dots dp_{3N}$, so this may be added to the integrand to obtain, finally

$$\bar{H}_1 - \bar{H}_2 = \int \dots \int \left\{\rho_2 \log \frac{\rho_2}{\mathrm{P}_2} + \mathrm{P}_2 - \rho_2\right\} dq_1 \dots dp_{3N}. \qquad (2.56)$$

The integrand in (2.56) is always positive, except when $\mathrm{P}_2 = \rho_2$ in which case it is zero. Thus

$$\bar{H}_1 > \bar{H}_2 \qquad (2.57)$$

except in equilibrium. (2.57) is the Γ space analogue of the single system H theorem (2.29).

In the next chapter the nature and relationship of the various assumptions A to D will be examined, and the conditions and validity of the various theorems described in this chapter clarified.

2.6 The meaning of entropy

The identification of Boltzmann's H function with negative entropy (2.31) opens the way to a much greater understanding of the significance of entropy than is provided by the thermodynamic definition. It has already been described how equilibrium is achieved by the redistribution of molecules from an arbitrary initial state to a more *probable* equilibrium state. The most probable macroscopic state is realizable by the greatest number of microstates, but it also has the greatest entropy. In fact, a detailed analysis reveals the relation

$$S = \mathrm{k} \log G + \text{constant} \qquad (2.58)$$

where G is the total number of microstates consistent with the given macrostate whose entropy is S (this being just the probability that the macrostate be achieved after a random 'shuffle' of molecules). G is related to the occupation numbers n_i for the cells i. Note that although the probabilities G are multiplicative are usual, the logarithm in (2.58) assures that the entropy S is an additive quantity.

In view of the fact that equilibrium is achieved by the constant 'shuffling' of microstates, it is expected that the maximum entropy state will be the most 'shuffled' state. Put another way, chaotic disordered states are more probable than structured ordered ones. (These remarks are not immediately applicable in the case of gravitating systems—see section 4.7.) Many familiar examples are available from thermodynamics to support this assertion. For example, if a membrane is removed from between two different gases enclosed in a vessel, the equilibrium state reached after a few moments would be one in which the gases were evenly mixed throughout the vessel. The advantage of the interpretation (2.58) is its possible application to non-thermodynamic systems. Thus, if the above experiment were performed with two sets of macroscopic elastic balls of different colours replacing the gas molecules, the entropy of the system could be defined in the same way, with the same 'mixing' result.

If the increase in entropy is identified with the removal of structure, as the above picture suggests, then we arrive at an important connection between entropy and *information*. A more structured state clearly requires more information to specify it macroscopically (a microscopic description of N particles requires a constant $6N$ parameters for a complete dynamical description irrespective of the macroscopic structure, or lack or it). The removal of structure then corresponds to the loss of information, leading to the interpretation of information as *negative entropy*.

To be more precise, consider a general situation in which there are a discrete number p_0 of possible outcomes. In the absence of any information about the situation, each outcome is equally likely. However, it may happen that we have some information which enables us to reduce the number of choices p_0 to $p < p_0$. In that case the quantity of information ΔI is defined to be

$$\Delta I = k \log(p_0/p) = -k \log p + \text{constant} \tag{2.59}$$

where k is a constant. In physical applications k is usually taken to be Boltzmann's constant $\sim 10^{-16}$ erg $°K^{-1}$. The smallness of this number is an expression of the fact that 'everyday' systems contain vast amounts of information. A comparison of equations (2.58) and (2.59) as applied to a box of gas reveals the connection between information and negative entropy. An appropriate choice for the constants leads to the precise relation

$$\Delta I = -\Delta S \tag{2.60}$$

where $-\Delta S$ is the *loss* in entropy of the gas when restricted by the additional information ΔI.

A practical example will illuminate equation (2.60) by showing how negentropy (negative entropy) and information may be transformed into one another. Suppose we wish to ascertain something of the condition of a gas in a flask. A possible method would be to pass a beam of light through the gas and observe the scattering and refraction which results. The price paid for this information is the unavoidable degradation of some low entropy light into heat due to absorption by the gas. This is an example of

$$\text{negentropy} \rightarrow \text{information.}$$

Conversely, if it is known that all the molecules of the gas are residing for a short while in only one half of the flask, then a partition may be inserted at the appropriate moment so as to trap the gas in the smaller volume indefinitely, thereby decreasing the total entropy. Therefore

$$\text{information} \rightarrow \text{negentropy}$$

may be regarded as the power of *intervention* in a physical system.

It is a general principle that no physical system may be observed without perturbing it in some way (cf. section 6.3 on quantum measurement theory). For example, finding the temperature of a gas requires the use of a thermometer which upsets the thermal equilibrium of the gas. The acquisition of information must always be paid for by the expenditure of a small but finite quantity of negentropy. This is expressed in the 'negentropy principle of information' due to Brillouin[13], which states that a gain in information ΔI must be accompanied by an increase in entropy ΔS of the system, such that

$$\Delta S \geqslant \Delta I. \tag{2.61}$$

This principle seems to preclude the possibility of an intelligence contriving to violate the second law of thermodynamics by sufficiently dexterious activity. Such was the proposal concerning the famous Maxwell demon[14], a hypothetical being of microscopic proportions capable of perceiving and redirecting single molecules. If such an individual were to be placed near a hole in a screen dividing two halves of a gas-filled container, he could operate a shutter mechanism in such a way that molecules would only be permitted to pass in one direction through the hole or, alternatively, fast moving molecules one way and slow moving the other. Eventually the entropy of the system would be reduced as the molecular redistribution established a pressure or temperature gradient between the two halves of the container.

Maxwell's demon was for many years a subject of controversy, but it was eventually shown by Szilard[15] and Brillouin[13] that the effective operation of the demon always created at least as much entropy as it destroyed. In order to operate the shutter at the appropriate moment the demon would need to know the positions and velocities of the molecules in his vicinity. To acquire this information he would have to disturb them in some way, for example by illuminating them. This would upset the thermal equilibrium of the container and so increase the total entropy. Inequality (2.61) requires that this entropy increase would be at least as much as the information gain upon which the demon could act to reduce the entropy.

Everyday experience indicates that information only *increases* with time. Our own memories grow as we do, public libraries accumulate books, the moon accumulates craters from meteoric impacts. There is no incompatibility between the simultaneous growth of both entropy and information, once it is realized that traces which are left behind by some event or circumstance may increase the information content (hence decrease the entropy) of some local region of the universe, but only at the expense of an entropy increase in the overall universe. The law of entropy increase refers to *closed* systems, the law of information increase to *open* systems.

References

1. See for example I. Prigogine, *Non-Equilibrium Statistical Mechanics*, Interscience, New York, 1962.
2. R. Clausius, *Phil. Mag.*, **14,** 108, 1857; **40,** 122, 1870.
3. J. C. Maxwell, *Phil. Mag.*, **19,** 19, 1860; **20,** 21, 1860; *Phil. Trans. Roy. Soc.*, **157,** 49, 1867. Reprinted in *The Scientific Papers of James Clerk Maxwell* (Ed. W. D. Niven), Dover, New York, 1965.
4. L. Boltzmann, *Wien. Ber.*, **53,** 195, 1866; **63,** 397, 1871; **66,** 275, 1872; *Lectures on Gas Theory 1896–1898* (English translation by S. G. Brush), University of California Press, Berkeley and Los Angeles, 1964.
5. See for example K. Huang, *Statistical Mechanics*, Wiley, New York, 1966, p. 82.
6. K. Huang, *Statistical Mechanics*, Wiley, New York, 1966, p. 64 gives a clear exposition.
7. R. C. Tolman, *The Principles of Statistical Mechanics*, Oxford University Press, London, 1938, sections 41 and 42.
8. J. W. Gibbs, *Elementary Principles in Statistical Mechanics*, Yale University Press, New Haven, 1902; *The Collected Works of J. W. Gibbs*, vols. 1 and 2, Yale University Press, New Haven, 1948.
9. G. D. Birkhoff, *Proc. Nat. Acad.*, **17,** 650; **17,** 656, 1931.
10. J. von Neumann, *Proc. Nat. Acad.*, **18,** 70; **18,** 263, 1932.
11. R. C. Tolman, *The Principles of Statistical Mechanics*, Oxford University Press, London, 1938, chapter 3.
12. P. and T. Ehrenfest, *The Conceptual Foundations of the Statistical Approach in Mechanics*, Cornell University Press, Ithaca, 1959.

13. L. Brillouin, *J. Appl. Phys.*, **22,** 334, 1951.
14. J. C. Maxwell, *Theory of Heat*, Longmans, Green & Co, London, 1908, p. 338.
15. L. Szilard, *Z. Phys.*, **53,** 840, 1929.

Further reading

1. I. E. Farquar, *Ergodic Theory in Statistical Mechanics*, Interscience, London, 1964.
2. D. ter Haar, *Elements of Statistical Mechanics*, Oxford University Press, London, 1938.
3. J. H. Jeans, *An Introduction to the Kinetic Theory of Gases*, Cambridge University Press, Cambridge, 1940.
4. L. D. Landau and E. M. Lifshitz, *Statistical Physics*, Pergamon Press, London, 1958.
5. R. C. Tolman, *The Principles of Statistical Mechanics*, Oxford University Press, London, 1938.
6. C. E. Shannon and W. Weaver, *The Mathematical Theory of Communication*, University of Illinois Press, Urbana, 1949. R. M. Lewis, *Science*, **71,** 569, 1930.

3 The Reversibility Objections

3.1 The symmetry of isolated systems

In section 2.3 we outlined Boltzmann's famous proof that the function H, defined by equation (2.24), can only decrease in time until it reaches its minimum value when the gas has a Maxwell velocity distribution, and is in equilibrium. The proof is based on the Stosszahlansatz, or assumption of molecular chaos, B. Some of the objections raised against the H theorem, and its underlying assumptions, will now be considered.

It was pointed out by Loschmidt[1] that the H theorem as stated must be incorrect, because it contradicts the principle of microreversibility on which it is based (equation 2.14). Because each individual collision is reversible, for every set of motions which decreases H, there will be a corresponding set which increases H, obtained from the first set by reversing all the particle velocities simultaneously. This 'time reversed' system would then pass backwards through a reversed sequence of states; H cannot decrease in both cases.

A second objection to the unrestricted H theorem was raised by Zermelo[2], based on a theorem due to Poincaré[3]. Known as Poincaré's *recurrence theorem*, it states that in an isolated system, *any* state will be revisited to arbitrary closeness an infinite number of times. The time between visits is known as the *recurrence time*, and is enormously long for macroscopic systems (say, $10^{10^{23}}$ seconds). Nevertheless, it is clear that *eventually* even very low entropy states will occur, so that H must decrease sometimes.

Poincaré's theorem may be proved as follows. The swarm of representative points in Γ space move along a tube of constant cross section. For finite systems with bounded energy, the points are restricted to a finite volume of Γ space; hence the total volume of the tube is finite. Consider the part of the tube in the future of region A in figure 3.1. Let it have volume Ω_A. This volume contains all the points in the tube to the future of region B, because A evolves

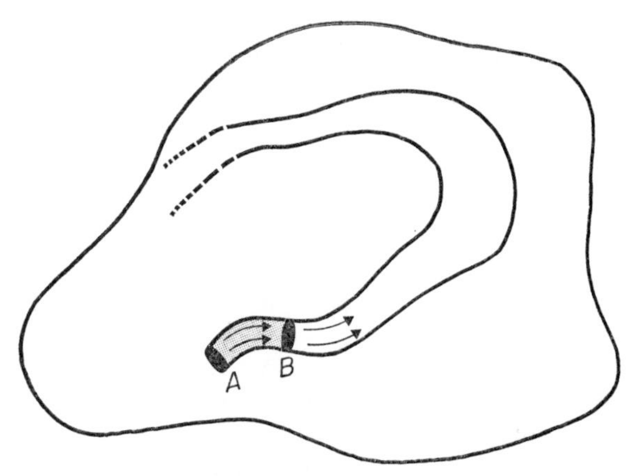

Figure 3.1 Poincaré's theorem. The future tube of region A evolves into the future tube of region B as the the representative points stream out of the shaded part of the tube between A and B. As all the points in A's future become all the points in B's future, A and B must have the same future tube. Hence the shaded region must also lie in the future of B as well as its past. This means that a point in A will always return to A.

into B after a time t. Let this latter volume be Ω_B. Imagine Ω_A to be filled with representative points. Then after a time t all these points will be in Ω_B. But Ω_A contains Ω_B, so this requires $\Omega_A = \Omega_B$. Therefore, the region of the tube between and including A and B (which certainly lies in Ω_A) must lie in Ω_B. It follows that the future tube of B contains all points in A (except for a set of measure zero). If a region A lies both in the past and future of B, a point in A must obviously return to A after a sufficiently long time. As A may be made arbitrarily small, any state will be revisited to arbitrary accuracy. The argument may then be repeated infinitely often.

We must now inquire as to how the H theorem, which apparently demonstrates that entropy can only increase (H decrease), may be reconciled with these remarks on reversibility and recurrence. The H theorem rests on the assumption B of molecular chaos, and it is not hard to see that this has been made asymmetric in time. It will be recalled that the assumption was interpreted in chapter 2 to mean that the positions and velocities of particles were uncorrelated *before* they collided, based on the supposition that a future collision could not influence the properties of particles that were about to collide. The distinction of 'before' and after (when the particles are correlated) is responsible for the apparent one way in time increase in the entropy.

The precise physical meaning of the assumption of molecular chaos is unclear. As it arose in equation (2.12), it enabled the number of collisions of a certain kind to be written in terms of a product of distribution functions

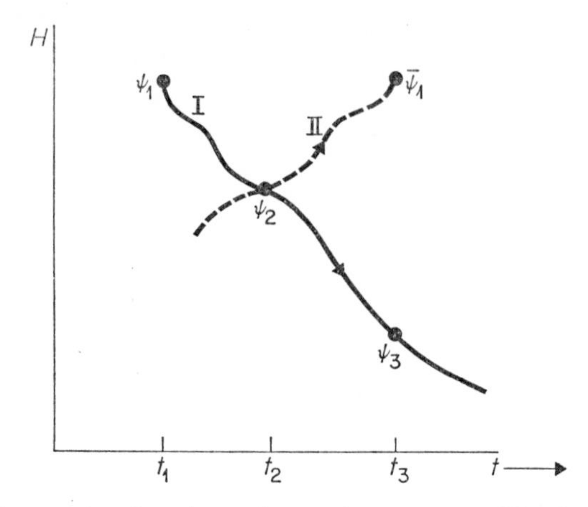

Figure 3.2 System I evolves down the continuous curve. The time reversed system II must therefore evolve *up* the mirror (broken) curve. If I and II have the same distribution function at t_2 then the H theorem predicts that H will decrease in both cases. The H theorem must be invalid; there cannot be molecular chaos at t_2.

$f_1 f_2$. However, the assumption places no restriction on the *form* of the distribution function; any particular f may be selected and molecular chaos imposed in addition. In the treatment of the H theorem given in section 2.3, assumption B was made at each point in time as H decreased. This cannot be correct, for it is easy to construct distribution functions for which such a situation would contradict the laws of mechanics[4]. As an example, consider a system called I, which passes through a sequence of states $\psi_1, \psi_2, \psi_3, \ldots$ at times $t_1, t_2, t_3, \ldots$ along a monotonically decreasing H curve (continuous line of figure 3.2). Consider another system II in a state $\bar{\psi}_2$, which is the same as ψ_2 except that all the molecular velocities are reversed. Suppose a distribution function is chosen for ψ_2 that depends only on the magnitude of the velocity: $f = f(|v|)$; then ψ_2 and $\bar{\psi}_2$ have the same distribution function, and hence the same H. If molecular chaos is imposed on ψ_2 and $\bar{\psi}_2$ then the H theorem rigorously requires that H decreases subsequently for *both* systems I and II. On the other hand, system II is the time reverse of I at time t_2. Therefore II evolves at later times in reverse order through the earlier ($t < t_2$) sequence of states of I, i.e. back long the continuous line in figure 3.2 towards ψ_1. The laws of mechanics thus require H to *increase* in system II. It follows that there can only be molecular chaos when H is at a local *maximum* (entropy minimum). Only in this case is the assumption compatible with the laws of mechanics.

Because molecular chaos can only be true at a single moment, corresponding to a peak in the value of H, the H theorem can no longer be used to prove the continuous approach of an isolated system to equilibrium.

Given that B is valid at a particular instant, molecular collisions will soon destroy the molecular chaos by introducing correlations between the molecules which have suffered collisions. Then it may be said only that $\frac{dH}{dt} < 0$ an infinitesimal time later.

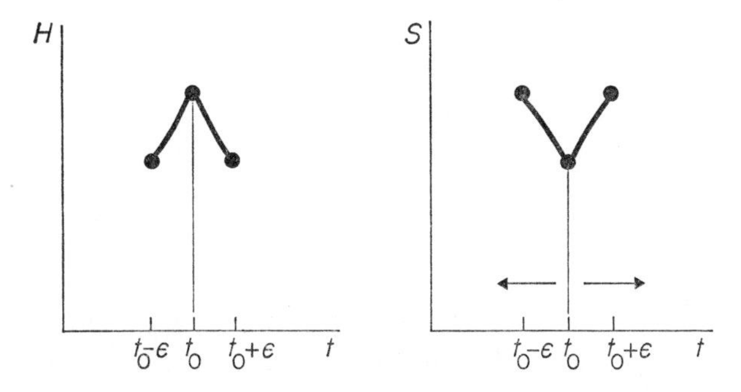

Figure 3.3 A manifestly time symmetric statement of the H theorem. Molecular chaos is valid only at the local peak where H decreases in both directions of time. On the right is drawn the corresponding entropy minimum. Boltzmann's equation may be applied to either ascending branch in a time sense *away* from t_0.

It is now possible to reconcile the H theorem with the time reversal properties of the particle motions. If a gas is in a state of molecular chaos at a time t_0, then the H theorem says $\frac{dH}{dt} < 0$ at $t_0 + \epsilon(\epsilon \to 0)$. However, the value of H may also be asked for at earlier times $t < t_0$. The argument is perfectly symmetric, so, of course, $\frac{dH}{dt} > 0$ at $t_0 - \epsilon$ (see figure 3.3).

The long time behaviour of the entropy of an isolated system is drawn in figure 3.4. The curve shows a succession of very small *fluctuations* about the maximum entropy equilibrium state. Occasionally there is a large fluctuation as the system revisits some low entropy state in accordance with the theorem of Poincaré. Molecular chaos applies at the bottom of such a fluctuation. The most important aspect about this picture from our point of view is its complete lack of temporal directionality; every fluctuation has two sides, a downgrade and an upgrade. This is true whichever way time is taken to increase on the horizontal axis; entropy decreases as often as it increases. It must be concluded that an isolated box of gas *cannot show an asymmetry in time.*

It is also clear from the foregoing remarks that there can be no true 'equilibrium' state for such a system. The state that we have called equilibrium is only the most frequented state, and does not satisfy the usual criterion of

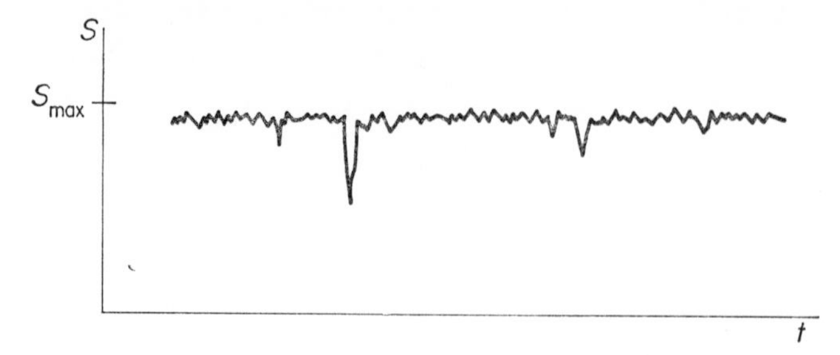

Figure 3.4 Long time behaviour of the entropy of an isolated system. There are vastly more small fluctuations than large. The system is obviously not oriented in time.

equilibrium because the system will leave it eventually (though only appreciably after vast periods of time). For this reason, all statements about equilibrium and irreversibility in this book should be interpreted as meaning on time scales much less than the Poincaré recurrence times.

3.2 The generalized $\bar{H}$ theorem again

Whilst the reversibility objections invalidate the use of the unrestricted H theorem of section 2.3, the generalized $\bar{H}$ theorem remains unaffected. The latter is not based on the false assumption that molecular chaos is always valid, but rests instead on the assumption D of equal *a priori* probabilities at the initial instant. Because this assumption D is only applied at one moment, it cannot contradict the reversible equations of motion. (Assumptions B and D are equivalent for a uniform gas at one instant.) The decrease in $\bar{H}$ is based on the fact that although ρ and P are assumed equal initially, they are not equal at later times. It appears very plausible (though quite without proof) that $\bar{H}$ will continue to decrease until the minimum value has been reached, corresponding to the microcanonical ensemble.

To picture how this comes about, suppose that the representative point of the system of interest is known to be located in some finite region of Γ space, corresponding to the particular macrostate observed. That is, there is a definite uncertainty about where the point is, and consequently about where it was in the past and will be in the future. In section 2.4 Liouville's theorem was proved, which states that if the region was filled with representative points they would move along a tube in Γ space with *constant* cross section. Consequently, it might be thought that the uncertainty about the position of the representative point of our single system, i.e. the uncertainty about the state of that system, would remain constant also, because any choice of point in the region near the true position would remain inside the tube.

However, nothing has been said about the *shape* of the tube cross section. In fact, because of the enormous number of collisions which occur in a realistic system, two points in Γ space initially close together will soon move very far apart, as even small initial differences will carry the two systems on widely different trajectories. Consequently, if we follow the development of a small hypersphere around the point of interest, the tube will not retain a spherical cross section but will alter its shape rapidly by growing a web of filaments as initially nearby points move away from one another. This situation is depicted in figure 3.5 which shows how the points in region *A* spread out into a filamentary structure at *B*, although the total volumes of *A* and *B* are the same by Liouville's theorem. Eventually these filaments wind their way throughout the whole of the energy surface so that the density of points becomes more and more uniform, converging towards the microcanonical ensemble as the system approaches equilibrium.

The relevance of this discussion to the asymmetry of a macroscopic description is now clear. Suppose that the true position of the point is at *P* in region *A*, but that it is guessed to be a small distance away at *Q*. After a certain time when the points in *A* have moved on to *B*, *P* and *Q* will have moved to *P'* and *Q'* respectively. However, because of the filamentary structure of *B*, the separation of *P'* and *Q'* has become very large. Because the details of the filamentary structure cannot be 'observed' macroscopically,

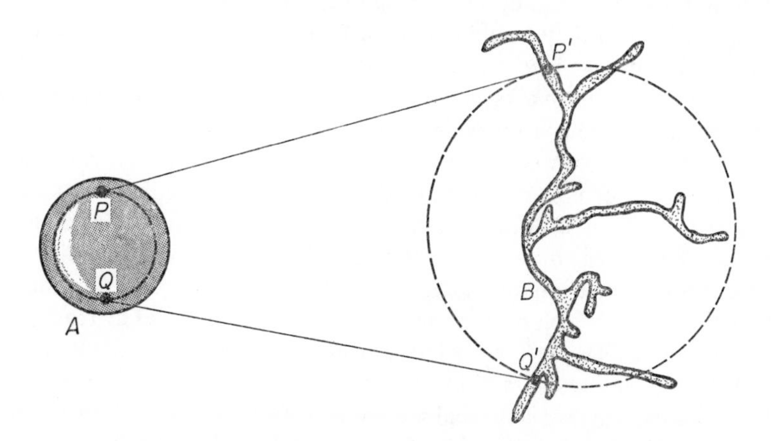

Figure 3.5 Points *P* and *Q* initially close together in the hypersphere *A* move apart as *A* evolves into the equal volume but filamentary structure *B*. If a macroscopic observation cannot distinguish *P* from *Q*, the region of uncertainty (broken circle) grows outwards with the filaments, until it encompasses the whole energy surface.

the region of uncertainty around P grows outwards with the filaments. We are therefore progressively less sure about our predictions for progressively later times. Because the loss of predictive power implies loss of information about the system, the spreading of the filaments implies a gain of coarse-grained entropy, in accordance with the discussion of section 2.6. If our observations are restricted to a limited period, the phenomenon just described appears to lead to a temporal asymmetry. In fact, the microcanonical ensemble would be expected to unmix itself after a sufficient time. There is an appropriate analogy due to Gibbs[5] which considers the effect of stirring a mixture of water and black ink. Practically any kind of stirring will soon produce a uniform grey colour when viewed on a macroscopic scale. However, microscopically, the fine-grained densities of the water and ink are unchanged; it is just that we are not capable of observing the fine drawn out filaments of the two fluids. In addition, given sufficient time, one might reasonably expect that the water and ink would unmix again, as chance arrangements of fluid tended to build up concentrations in one region. Thus Poincaré type cycles of $\bar{H}$ might be expected also, as exceedingly rare fluctuations in the microcanonical ensemble take place. Such a recurrence theorem cannot be proved, because the ensemble is a continuum of points, unlike the molecules of a gas[6].

It is important to note the crucial role played in the generalized $\bar{H}$ theorem by the introduction of coarse graining (finite sized stars in Γ space). Each star is filled with system points, some of which represent individual systems that do not approach equilibrium owing to the peculiar nature of their initial conditions. Instead they would execute a Poincaré type fluctuation to lower values of the entropy. However, such systems are exceedingly rare compared to the many other 'normal' systems occupying a star, so that the appearance of abnormal behaviour is exceedingly improbable, though not impossible.

In conclusion, it can be said that the generalized $\bar{H}$ theorem contributes qualitatively to our understanding of the approach to equilibrium. It shows that, starting with an arbitrary value of $\bar{H}$ at an initial instant, it is at a lower value at a later time. It also appears reasonable that the value continues to decrease until the coarse-grained density reaches a stationary (microcanonical) distribution, after which no further decrease can occur as it has reached its minimum value. The last assertion can never be proved without dealing directly with equations of motion, i.e. integrating Liouville's equation, which is the very situation that statistical mechanics sets out to avoid. In particular, the $\bar{H}$ theorem gives no indication of the *time* required for equilibrium to be reached. For a better description of the approach to equilibrium, it is necessary to return to the single system H theorem of Boltzmann, but to modify it by adopting a truly *statistical* treatment that is free of the reversibility objections.

3.3 The statistical single system H theorem

In the previous section it was seen how the generalized $\bar{H}$ theorem contributes only qualitatively to our understanding of the approach to equilibrium in an isolated system. To follow in detail the decrease of $\bar{H}$ to its minimum value, we should need to know ρ, and thus P, at each instant, which can only be obtained by an exact integration of the equations of motion. Thus a rigorous proof of the approach to equilibrium is lacking.

In order to remedy some of the inadequacies of the ensemble treatment, without actually integrating the equations of motion, we return now to the individual physical system, and briefly describe the statistical version of Boltzmann's H theorem, which is free of the reversibility objections of Loschmidt and Zermelo.

Attention is now focussed on a single representative point in Γ space corresponding to the single system of interest. At any moment it may only be said with certainty that the point lies in a given star Λ of finite volume $W(\Lambda)$. H is defined by equation (2.24) or alternatively by the equivalent definition

$$H = \sum_i n_i \log n_i + \text{constant} \tag{3.1}$$

with

$$N = \sum_i n_i \tag{3.2}$$

in terms of the occupation numbers n_i of the cells i in μ space. Definition (3.1) emphasizes that H really changes in discrete jumps as the system point moves from star to another. As mentioned in section 2.3, H is minimized by the Maxwell distribution, and it is also this distribution that is associated with the *largest* star in Γ space. Indeed, the volume $W(\Lambda)$ of the star corresponding to the equilibrium distribution is overwhelmingly larger than any other volume. Because of the postulate of equal *a priori* probabilities, it follows that the Maxwell distribution is the most probable distribution for an isolated box of gas inspected at random. Moreover, if the representative point is ergodic, and spends equal intervals of time in equal volumes of Γ space, it will spend the overwhelming fraction of its time in the equilibrium state. Sometimes it will make a transition to another star corresponding to a value of H slightly larger than the minimum; the gas will then execute a fluctuation. It is then far more likely that the point will rapidly return to the equilibrium star, rather than make a further transition to a star of still smaller volume, corresponding to a still higher value of H (lower value of S). Eventually, however, all stars will be visited, whatever their volumes.

This motion of the representative point among the various stars is of course perfectly reversible, because of the time symmetry of the equations of motion (2.32) and (2.33). In order to establish an H theorem which does not contradict these reversible equations, without actually having to solve them, we proceed as follows. Assume initially ($t = t_0$) the point resides in a particular non-equilibrium star. Try to describe its *most probable* behaviour, as it makes transitions from one star to the next, presumably almost always in the direction of the largest (equilibrium) star.

Differentiating equation (3.1) and using (3.2) gives

$$\frac{dH}{dt} = \sum_i (1 + \log n_i) \frac{dn_i}{dt} = \sum_i (\log n_i) \frac{dn_i}{dt}. \tag{3.3}$$

Hence the behaviour of H is determined by the quantities $\frac{dn_i}{dt}$, which are related to the transition probabilities for the point to pass from one star to the next. These transition probabilities may be calculated using a representative ensemble chosen in accordance with the fundamental assumption D. This therefore amounts to a *statistical* treatment of the one system H function. We are now concerned with the most *probable* value of $\frac{dn_i}{dt}$, and hence the most probable value of $\frac{dH}{dt}$, based on assumption D which replaces assumption B of the exact treatment (section 2.3).

For the details of the proof of this statistical H theorem, the reader is referred to Tolman[7]. It follows closely the treatment of section 2.3, with the same restriction to binary collisions. Rather than the exact calculation of $\left(\frac{\partial f}{\partial t}\right)_{\text{coll}}$, a quantity $a_{ij\to kl}$ is now introduced, denoting the most probable (i.e. ensemble average) number of points per unit time to move as a result of collision out of cells i, j into cells k, l. $a_{ij\to kl}$ may clearly be written in the form

$$a_{ij\to kl} = A_{ij;kl} n_i n_j \tag{3.4}$$

where $A_{ij;kl}$ depends on the structure of the collision potential. Actually, for a box of gas with volume V

$$A_{ij;kl} = \frac{|\boldsymbol{v}_i - \boldsymbol{v}_j|\, \sigma(\boldsymbol{v}_i, \boldsymbol{v}_j; \boldsymbol{v}_k \boldsymbol{v}_l)\, d\Omega}{V} \tag{3.5}$$

(cf. expression (2.12)). The principle of microreversibility states the equality of $A_{ij;kl}$ with the corresponding quantity for the reverse process, which for spherical particles is the same as the inverse

$$A_{ij;kl} = A_{kl;ij}. \tag{3.6}$$

Using these results in equation (3.3), it is easily seen that the most probable value of $\frac{dH}{dt}$ is negative

$$\left(\frac{dH}{dt}\right)_{prob} < 0. \tag{3.7}$$

The result (3.7) is only valid for macroscopically short times after t_0, for it is only at the initial moment t_0 that the representative ensemble, constructed in accordance with assumption D, is valid. ρ and P will not in general be equal at a later time.

The statistical statement of the H theorem (3.7) is quite consistent with the reversibility and recurrence paradoxes. It is a statement about the *most probable* behaviour of a non-equilibrium system, calculated using an ensemble which contains overwhelmingly points representing systems which *do* subsequently decrease the value of H. However, the ensemble inevitably contains other rare examples of systems which increase their H value. These cases are included in the ensemble because a specification of the distribution function f, or the equivalent occupation numbers n_i, is a macroscopic requirement, and thus consistent with the rare, but possible, situation in which all the particles happen to be poised on the brink of a Poincaré type fluctuation. As the microreversibility principle and recurrence theorem refer to exact results of an *individual* system, they are not in contradiction with a statistical statement about *many* similar systems. The fact that after a sufficient period of time a *given* system will undergo entropy decreases, is quite consistent with the claim that *if* a system has a low value for the entropy, it is vastly more probable that this entropy will subsequently increase than decrease. Put another way, if an isolated system is encountered in a low entropy state, it is overwhelmingly probable that the assumption of molecular chaos is correct, though in very rare cases it will not be so, and then the unrestricted H theorem will not apply.

It is possible to relate the statistical one system H theorem of this section with the generalized $\bar{H}$ theorem of section 2.5. If the H of the present section is denoted by H_{syst} to indicate that it refers to the individual system, then it follows from the definitions of $\bar{H}$ and H_{syst} that

$$\bar{H} = \bar{H}_{syst} + \sum_{n} \mathrm{P}_n \log \mathrm{P}_n \tag{3.8}$$

where $\bar{H}_{syst}$ is the ensemble average of Boltzmann's entropy H_{syst}, and the second term is a sum over all the stars n of quantities which describe the distribution of the ensemble over the stars[8]. Differentiating (3.8)

$$\frac{d\bar{H}}{dt} = \frac{d\bar{H}_{syst}}{dt} + \frac{dP_n}{dt} \log P_n \tag{3.9}$$

shows that the decrease in $\bar{H}$ arises from two causes: (1) the decrease of the Boltzmann entropy for each individual system, (2) the spreading of the different system points in the ensemble more uniformly over the different stars n.

This section proceeds with a brief discussion on the possible validity of the H theorem beyond a small interval of time after the initial instant t_0[9]. The statistical interpretation (3.7) of the H theorem states that if the representative point of the single system is initially in a star for which H is not a minimum, it will very probably start to move towards a star with a lower value of H. If it is accepted that most of the time the point is in the star with the lowest value of H, it appears reasonable that a series of transitions from one star to another will take place until the point is back in its most frequented star. However, this is unproved unless the additional assumption is made of the conservation in time of the property of molecular chaos:

assumption B is true at every instant. (E)

It is now permissible to assert that (3.7) is valid from the initial instant t_0 until the system reaches equilibrium. In accordance with section 2.2 assumption E allows us to recover the time asymmetric Boltzmann equation (2.10).

Assumption E is obviously in contradiction with the laws of mechanics. It permits a new representative ensemble to be constructed in accordance with equal *a priori* probabilities at every instant of time. However, the microscopic evolution of the system will, in fact, relate the ensembles at successive moments. For example, if an observation at t_0 establishes that the representative point resides in a star Λ_0, and at a later time t_1 the point is found in another star Λ_1, we are not really free to construct a new ensemble based on the later observation in accordance with D, for we should have to remove from this ensemble all those system points that had not previously, at time t_0, resided in the star Λ_0. But these points can only be selected after an exact solution of the equations of motion has been performed. From the statement of assumption B, and in accordance with the discussion of section 3.1 concerning the conditions of its validity, it will be appreciated that the existence of collisions create correlations which destroy molecular chaos as soon as it has been established. The assumption E of the constancy of molecular chaos thus amounts to neglecting these correlations, i.e. assuming that the macroscopic state of the system has no 'memory'. The progressive 'loss of memory' (or information) is naturally associated with an increase in entropy. Assumption E therefore introduces a statistical element into the evolution of the system itself, and no longer just at the initial instant. Such an assumption can ultimately be justified only by the success of the Boltzmann equation in practical calculations, although some authors[17] attempt to justify this assumption by invoking the coupling of the system to the outside world through the walls of the container (see section 3.5).

Once it is accepted that the evolutionary path of the representative point in Γ space is described quite accurately by a statistical process, it is instructive to formulate the behaviour as a Markovian stochastic process[10]. The point may be regarded as performing a random walk through the different stars of Γ space; it then follows immediately that the system will almost always tend to equilibrium from an arbitrary state, because of the overwhelming fraction of the volume occupied by the equilibrium star. This random walk assumption is, of course, equivalent to assumption A—the random redistribution of points in μ space.

It should now be thoroughly appreciated by the reader that an isolated box of gas cannot be asymmetric in time.

If t is chosen, quite arbitrarily, to increase to the right in figure 3.3, then the form of the H theorem stated in (3.7) is correct. It only applies to the right-hand branch of the figure. If t is taken to increase to the left, the H theorem applies instead to the left-hand branch. In either case, a corresponding 'anti-H theorem' with a minus sign in (3.7) must be introduced to describe the other branch; there is complete symmetry. Of course, an 'anti-Boltzmann equation' is also required, with a reversed sign for the $\frac{\partial f}{\partial t}$ term, to apply to this opposite branch[11].

Since the Boltzmann equation is only strictly valid at the instants of molecular chaos, it can only approximately describe the approach to equilibrium along one or other of the branches in figure 3.3, because the collisions progressively destroy the assumed molecular chaos. (Of course, there might be many small scale dips in the ascending entropy curve. Molecular chaos will then apply at the bottom of these dips[12].) For this reason the equation cannot be inverted and used to predict the growth of a fluctuation—the descent of the entropy from a maximum value into a valley. It can only be used to describe the behaviour *away* from molecular chaos (in the direction of the arrows).

Now the Boltzmann equation is a parabolic differential equation which requires specification of boundary conditions for its solution. From the remarks in the last paragraph we are obliged to take this boundary condition at the time of mimimum entropy (at the bottom of the valley). As the equation is a macroscopic equation, the boundary conditions are in the form of macroscopic statements.

This may be illustrated with an example. Rather than discuss the Boltzmann equation as such, the heat conduction equation will be considered. This is also a parabolic equation which only applies to entropy increasing situations, and will serve equally well for our purposes here. Suppose that we have two blocks of metal in contact at different uniform temperatures. The heat conduction equation will predict the subsequent redistribution of heat energy until the system approaches (exponentially) a uniform temperature. It is overwhelmingly probable that this prediction is correct, although only a macroscopic condition—the temperature distribution—was specified at $t = 0$. If we wished to reverse the procedure, and describe the situation going back into the past from a uniform temperature distribution to the low entropy condition where the blocks were at different temperatures, the heat conduction equation could not be used. In fact, *any* final distribution of temperature (including an already uniform one) would produce a uniform distribution at sufficiently early times. The desired retroduction could only be carried out with the precise *microscopic* knowledge of all the particles in the blocks.

This asymmetry between prediction and retrodiction[13] can be easily understood using the informational interpretation of entropy mentioned in section 2.6. In an entropy increasing situation, information is progressively lost, so that a macroscopic description of the boundary conditions is sufficient. However, in an entropy decreasing situation, information progressively increases, so that *more* information than the macrostate will provide is required, i.e. a microscopic specification is necessary. Notice that there is still no asymmetry in time about these remarks concerning prediction and retrodiction. *Prediction* is used to discuss how a system evolves from an entropy minimum, along either of the branches of figure 3.3, to a maximum. It is correlated with entropy change, not time. Thus, in the example of the two blocks just discussed, nothing was said about how they reached their non-uniform temperature originally. For an isolated system this could only have come about by a fluctuation. During the growth of this fluctuation the direction in time of 'prediction' is reversed. (Some authors[13] prefer to fix an arrow pointing to the right on the time axis of figure 3.3 and then say that 'prediction' applies correctly to positive gradients and 'retrodiction' to negative gradients. Thus these operations are correlated to time direction rather than entropy change. Here the suffix 'pre' means 'before entropy increases', whichever way in time that is.)

3.4 Time asymmetry through branch systems

In the last section it was demonstrated that a finite, isolated box of gas could not display an asymmetry in time. Nevertheless, the asymmetry in time of the world about us is one of the most fundamental aspects of our experience, and it is necessary to inquire about the origin of this asymmetry, in the light of the considerations so far.

There are many ways in which the simple box of gas model explored by Boltzmann is an oversimplification of real, everyday thermodynamic systems. It could be argued, for instance, that real systems are not dilute, monatomic gases without many particle collisions, etc. However, it is hard to imagine that a reapplication of the Boltzmann analysis to ever more complex systems, including matter in many phases, would produce results essentially at variance with the simple case.

Far more relevant is the whole question of *isolation*. (1) It is well known that it is not possible to produce in the laboratory a perfectly isolated thermodynamic system. The most that can be hoped for is a *quasi-isolated* system, where the typical relaxation times for the system processes are much shorter than the time required for outside influences to cause appreciable disturbance. (2) In addition, even if it were possible to produce a truly isolated system, it could not be isolated *for all time*, for we should have made it ourselves; or, if it arose naturally, it would still be younger than the age of the universe.

The implications of remark (1) of the previous paragraph will be examined in the next section. This section will concentrate on remark (2). That is, our attention will be applied to the formation of systems which are quasi-isolated, on the justification that the small residual coupling to the outside world which is always present will not change the essential nature of the results, as will be seen.

It will be obvious to the reader that all our observations and experiments in thermodynamics refer to what Riechenbach[14] calls *branch systems*, rather than permanently isolated systems. Branch systems are regions of the world which separate off from the main environment and exist thereafter as quasi-isolated systems, and usually merge once again with the wider environment after a sufficient time. Examples of this sort are countless, but one will suffice to remove any misunderstanding. When we take an ice cube, add it to a lukewarm drink, and watch the ice cube melt, the system ice + drink only comes into being *after* this event. It simply did not exist as a quasi-closed system beforehand. Also it will be seen that it really is a quasi-closed system in the sense that the melting process can be perfectly adequately described without recourse to interaction with the outside universe.

Return for a moment to the case of a permanently, totally isolated system, and imagine that we have a piece of apparatus which is designed to make a signal when the total entropy inside a box takes on a certain value, which is supposed to be less than the maximum. If this value is rather close to the maximum it may happen that before very long a signal is observed. On the other hand, if it is far from the maximum we should have to wait for a very rare large fluctuation to trip the mechanism.

The following assertion is now made: when a signal is received, it is overwhelmingly probable that the gas is in a state of molecular chaos, i.e. at a local entropy minimum.

The basis of this assertion is the simple fact that there are vastly more small fluctuations than large ones, so that it is far more probable that the mechanism is tripped near the extremity of a small dip than on the continuing slope of a large one (see figure 3.4). In other words, if a low entropy state of the system is selected *at random*, it is almost always an entropy minimum, so that the entropy will almost always *increase* afterwards, whichever before–after direction is chosen on the time axis.

The relevance of this digression to the hypothesis of branch systems is as follows. The reason why a single permanently isolated system like that considered above cannot supply us with an asymmetry is precisely the fact that from such a randomly selected low entropy state, the entropy nearly always increases both in the past and future, as in figure 3.3. However, if a branch system is *formed* in a random low entropy state, it simply *did not exist* in the 'past' for the entropy to increase that way. This is indicated in figure 3.6, which shows the entropic behaviour of a branch system formed with entropy $S_0 < S_{max}$ at $t = t_0$. In accordance with the H theorem the entropy increases steadily to the right of t_0, but to the left of t_0 the branch system did not exist. So the state with entropy value S_0 can no longer be regarded as having formed from a random fluctuation from equilibrium (as would be described by the broken line). Instead it has been formed by *interaction with the outside world.*

A couple of examples will clarify this point. If we encounter a small ball near the rim of an upright hemispherical bowl we would be sure that a short while later the ball would be at rest at the centre of the bowl (see figure 3.6). However, if the system ball + bowl was for ever sealed in a perfectly insulated box, it would have to be concluded that the ball reached this position near the rim by the fortuitous fluctuation of large numbers of atoms in the bowl to produce 'negative friction' (see section 1.6), so accelerating the ball spontaneously to its low entropy condition at the rim. In other words, it would be concluded that the entropy-increasing frictional damping of the ball's motion *after* the instant of observation was mirrored by a corresponding entropy-decreasing 'anti-damping' *before* the observation. The reason why this conclusion is not usually drawn, and such unconventional behaviour not ascribed to the ball, is just because the system is *not* completely isolated. In fact, if we did encounter the ball in the rim position we would conclude that it had fallen or been dropped in that position from *outside*, so that the system ball + bowl only came into being immediately prior to that moment.

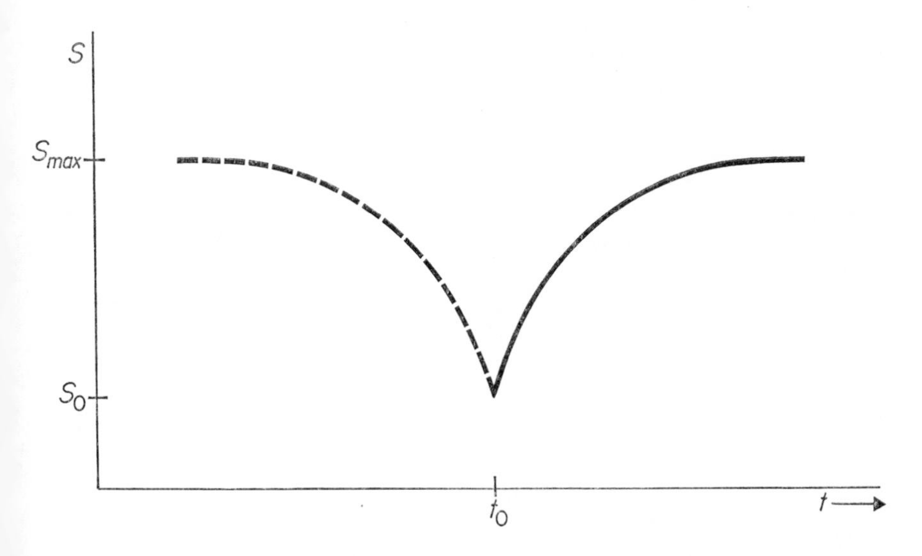

Figure 3.6 The solid line shows the entropy of a branch system formed at random at time t_0. The line is not continued to the left of t_0 because the system was not formed then. Consequently, t may be unambiguously chosen to increase to the right. In contrast, a permanently isolated system would almost certainly have reached the low entropy state S_0 at t_0 by a chance fluctuation from equilibirum, depicted by the broken line. In that case the entropic behaviour is clearly time symmetric.

Another famous example[15] concerns the conclusions of an individual who encounters a footprint on a lonely beach. If that part of the beach and its immediate environment was an isolated system, he would be forced to conclude that the imprint had been formed by the chance movement of sand particles by the wind and sea—an overwhelmingly rare occurrence. However, in practice, he would instead conclude that the beach was *not* an isolated system at all, but a branch system, the imprint having been formed from a recent encounter with a human stroller who, by his contact with the sand particles, coupled the system to the outside world.

It has been asserted by Reichenbach[14] that the entropic behaviour of branch systems supplies an unambiguous asymmetry to thermodynamics, which asymmetry is absent from a single permanently isolated system. The argument is summarized in figure 3.7, which shows the entropy curve of the main system, taken to be our local region of the universe, with a large number of branch systems separating off at points 1, 2, 3, ... These branch systems may be considered to be *completely* isolated from their moment of formation, i.e. the residual coupling with the main system, which is certainly present in practice, is ignored. If all the branch systems are formed in *random* low entropy states, then it has been seen already that the entropy of these new isolated systems will almost certainly increase, independent of our convention about the direction of increase of time. That is, it may be asserted confidently that almost all branch systems will show *parallel* entropy change. It is the asymmetry regarding the formation of the branch systems which brings about this parallel increase in all (nearly) branch system entropies. This asymmetry

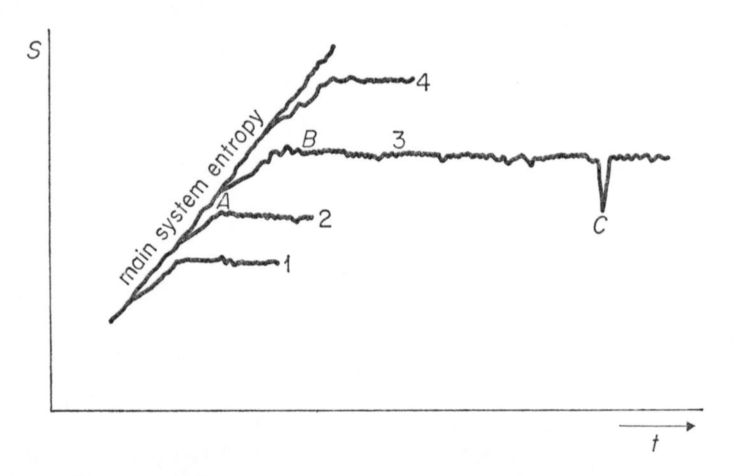

Figure 3.7 The overwhelming majority of branch systems 1, 2, 3, ... increase their entropy after formation (even if totally isolated thereafter). The chance of an entropy decreasing fluctuation, such as that shown in the time extended behaviour of system 3, is exceedingly small for times after the formation which are very much less than Poincaré recurrence times.

is definitely *not* supplied by statistics, coarse graining, the *H* theorem or anything else. If the branch systems are regarded as not existing in the *past*, then the entropy of the overwhelming majority of these systems will *increase with time*. It is through branch systems that the customary intuitive notion that entropy increases with time is derived.

With these considerations an entropy law can be formulated which is valid even in the case of a totally permanently isolated system. For if there is an ensemble of such systems which on a particular occasion are in randomly selected states of entropy $S < S_{max}$, then from the foregoing arguments it is clearly overwhelmingly probable that all these systems will be in molecular chaos at an entropy minimum, and so all will increase with both greater and lesser t, i.e. in both directions of time. If any two systems A and B are selected, and their entropies at the first mentioned occasion called S_A^1 and S_B^1 respectively, and on another occasion (not too widely separated from the first) called S_A^2 and S_B^2, then the product

$$(S_A^2 - S_A^1)(S_B^2 - S_B^1) \geqslant 0 \tag{3.10}$$

whichever way t is chosen to increase, for both A and B will almost certainly change in a parallel direction as seen. If $S_A^2 > S_A^1$ then $S_B^2 > S_B^1$, but if time is measured the other way and $S_A^2 < S_A^1$ so $S_B^2 < S_B^1$ and the product in (3.10) is positive in both cases. This is Schrödinger's[16] formulation of the entropy law, and it reduces to the usual form if A, say, is allowed to become the system of interest, and B become the rest of the world. Then if the increasing direction of time is *defined* as the direction in which the entropy of the whole world increases, (3.10) becomes

$$\Delta S \geqslant 0 \tag{3.11}$$

where $\Delta S = S_A^2 - S_A^1$. Equation (3.11) is just the law of Clausius (2.3).

This is confirmation of what is normally regarded as obvious, that the direction of entropy change in the branch systems is parallel to that of the outside world (the main system). In any case, whenever a branch system is formed in a state of low entropy by interaction with the outside world, and subsequently increases its entropy, this interaction itself will bring about an entropy *increase* in the total system. For example, the entropy decrease on the lonely beach due to the appearance of a footprint is more than compensated by an entropy increase in the surrounding world by the dissipation of energy in the muscles of the stroller, the heat flow into the surrounding regions as a result of friction, etc.

Branch system 3 of figure 3.7 is shown with time extended behaviour. The entropy of this system, as discussed, rises from A to B, at which it would be claimed that the system had reached 'equilibrium' However, being completely isolated, it would eventually (after a Poincaré cycle) return at C to a condition close to that at A, and we could no longer derive an asymmetry of time from the behaviour of the branch systems. In practice, of course, there would be appreciable disturbance of the system by the outside world before this time, and some branch systems may even cease to exist, becoming merged once again with the wider environment. Nevertheless, during periods of time small compared with the Poincaré recurrence times, when overall thermodynamic disequilibrium ensures that branch systems are still being formed in random low entropy states, their behaviour does provide a decisive, unambiguous asymmetry in time. It is clear also that *random* influences that might intrude on the system from the outside, as a result of the system not being completely isolated, will not change this result. That is, the asymmetry is *already present* even for the formation of totally isolated branch systems, so that we do not need to invoke the further, continuous coupling with the outside world that is always present thereafter to explain this asymmetry. As a rule, this latter type of coupling will have a negligible effect on the *macroscopic* entropic behaviour of the branch system during the increase from A to B, though it certainly destroys the *microscopic* reversibility of the system as will be seen in the next section.

3.5 The influence of the outside world

That the entropic behaviour of branch systems is essentially unchanged by the continuous interaction with the outside world always present is confirmed by many familiar examples from everyday life. To take an example already discussed, it is clear that an ice cube placed in a luke warm drink will melt whether or not minute random influences from the rest of the galaxy are disturbing the molecules therein. There is no doubt that a clock will run down if placed in a sealed box, whether the effect of the motion of the walls of the box on the clock are taken into account.

Certainly, these influences will contribute to the general entropy increase in the system, because they destroy information (see section 2.6), but they are in no way necessary for this increase. They are, therefore, in no way *necessary* for the general asymmetry in thermodynamics. In addition, it may be assumed that they are not *sufficient*, for if they are truly random influences—a not unreasonable hypothesis—then they have no time asymmetry (the reverse of a random sequence is still a random sequence, and one such sequence is as good as another here). Nevertheless, the continual interaction of a closed system with the outside world has been invoked by a number of authors[17] as an alternative to coarse graining to produce an approach to equilibrium (though not an asymmetry in time!). This will now be examined.

There appears to be no way to totally isolate a system from its environment. Certain substances are very effective at preventing the flow of heat and matter, but there is no known substance or device that in fact appears to be capable of providing complete insulation. In this respect, the Boltzmann model of a gas confined in a vessel with rigid, elastic walls is a total fiction. Any real wall will be made of atoms (or at best a magnetic field, which is not impervious to electromagnetic radiation) and these atoms will be in a state of motion because of their own coupling with the outside world through the thickness of the wall.

A fundamental assumption used in deriving the statistical mechanical results of chapter 2 was the existence of a total Hamiltonian for the system, *including the interaction with the walls*. For rigid elastic walls a potential energy term may be added to $\mathscr{H}$ which has the form of placing the particles in an infinitely deep potential well. The existence of $\mathscr{H}$ permits the application of canonical equations of motion leading to the Liouville theorem, and subsequently the H theorem.

However, if the walls are not rigid, but in motion in some uncertain way, then we cannot write down a Hamiltonian and canonical methods cannot be used. Of course, if the atoms of the walls were made part of the description, it could be argued that this objection is invalid, but then the matter outside the walls would have to be considered, and so on. The only truly canonical system is the whole universe (and even then only if it is finite).

The justification for neglecting the motion of the wall atoms so far has been the fact that a perfectly isolated system already relaxes to equilibrium on its own, in a time much shorter than the time required for a gas to reach equilibrium with the walls of its container. It is the internal collisions of the gas molecules which cause the approach to equilibrium, not the collisions with the walls. Detailed calculations[17] show that the internal relaxation time —the time required for the gas to reach internal equilibrium—is typically 10^4 times shorter than the time required to reach equilibrium with the walls. This is not to say that the walls cannot contribute considerably to the behaviour of the gas in some circumstances. It merely affirms that the asymmetric behaviour of thermodynamic systems is already present in the formation of totally isolated branch systems, without the additional necessity of introducing interactions with the walls of the container.

However, it was mentioned in section 3.1 that for a totally isolated system there was no 'true' equilibrium state. Indeed, the very definition of equilibrium depends essentially on the restriction to *coarse-grained observations*; the equilibrium state being that macrostate which can be achieved by the greatest number of microstates (or which occupies the largest volume of Γ space). Some authors[17] have objected strongly to the use of coarse-graining techniques in the discussion of relaxation to equilibrium. The argument employed against coarse graining is that it is based upon assumed human limitations of observation that may well be superseded one day, and indeed already have been in one significant case[18]. The equilibrium discussed in chapter 2 is referred to by these authors as 'pseudo-equilibrium'. They argue that it does not represent an objective property of the gas, but only a subjective inability to distinguish different microscopic configurations. Protagonists of this point of view indicate that a gas released from an arbitrary state still retains all the complicated correlations between the molecules long after it 'appears' to us to have reached 'equilibrium'. A reversal of the molecular velocities would certainly carry the system back to its initial state (time reversed) again, proving that it had not 'forgotten' the initial conditions. The conventional counter is that such a reversal cannot in practice be effected (in the words of Boltzmann: 'Go ahead, reverse them'). It must be admitted that this appears somewhat evasive, and is actually invalidated in the case of the spin-echo experiments of Hahn[18] that indeed bring about a kind of reversal.

From a mathematical point of view, it is incontestible that the fine-grained

$\bar{H}$ function σ (2.47) is constant in time (2.49), and so does not possess any asymmetry properties. Moreover, the continued decrease in the coarse-grained $\bar{H}$ function, as described by the H theorems, is clearly due to assumption E, which introduces an entropy increase precisely because this allows information about correlations to be thrown away. The time asymmetry accompanying the formation of low entropy branch systems, therefore, undeniably rests on the questionable premise that this information is irrelevant or unobservable, an apparently subjective criterion consequent upon the macroscopic nature of the human world.

On the other hand, the random perturbations of the system walls are responsible for an objective asymmetry because they bring about a continued increase in even fine-grained entropy by progressively destroying information about the initial state. Only when the initial conditions have been completely forgotten may the system be considered to be in 'true' equilibrium. When this state has been reached, a reversal of molecular velocities would not be expected to reproduce the features of the initial state, but only retain the equilibrium. (This is an example of time asymmetry with microscopic irreversibility.) In general, the time required for this 'true' equilibrium to be established is much longer than that which is customarily regarded as the coarse-grained relaxation time.

The apparent conflict between the random perturbation and the coarse-graining attitudes to relaxation and equilibrium is removed when closer inquiry is made into the problems that each method is trying to solve.

First we must decide whether to attach significance to *collective* properties of physical systems. For example, if it is felt that a pendulum is a legitimate subject for study in physics, we are forced to conclude that such an object when set into motion will suffer damping from frictional effects, so that the amplitude of its oscillations is progressively diminished. Provided we restrict ourselves to normal intervals of time which are very much less than the system Poincaré recurrence time, it has to be admitted that the *pendulum*, as an entity, has behaved asymmetrically in time. Although in thermodynamic language this damping would be called a macroscopically irreversible change, there is still complete microscopic reversibility (when the surroundings are also taken into account). There is asymmetry, but (micro)reversibility.

In contrast to this, one could choose to consider the complete atomic structure of the pendulum and its surroundings, in which case the 'pendulum', as such, no longer has any significance; we are simply dealing with a collection of particles that behave entirely without any asymmetry (if the outside world is neglected). In this situation it is more appropriate to direct inquiries not to the problem of *asymmetry*, but to the question of *irreversibility* in the microscopic sense. Indeed, there is microscopic irreversibility in nature if the truly randon nature of the intruding influences from the outside world is accepted. But attempt should not now be made to use the undeniable existence of this micro-irreversibility to discuss the question of asymmetry on a macroscale. Of course, a finite system of particles will display a microscopic asymmetry in time, as the fine-grained entropy increases and the initial conditions become obliterated. This fact though, is only relevant to physicists performing spin-echo experiments, etc. It is of no significance to the clock-maker who wants to know whether his pendulum will generate heat and stop. It is indeed a matter of philosophy rather than physics to decide if the coarse-grained asymmetry is 'real' or not. In this connection it must be remarked that the damping of a pendulum to rest would normally be considered as an objective fact, and not a consequence of human limitations vis-à-vis Maxwell demons. This is even more so in the light of considerations in chapters 4 and 7 wherein the motion of the whole universe is found to depend on macroscopic damping effects. Two comments are then prompted: (1) there *is* no outside world in that case to exert random influences, (2) the motion of the universe is presumably objective.

A similar situation arises in the subject of biology. The fact that the individual atoms of the human body are not themselves alive does not remove the significance of a person. The collective system is still a legitimate subject for study, having its own laws and phenomena, e.g. conditioned reflexes, which seem to be no less real for the fact that they are 'really' caused by the motions of individual inanimate atoms.

Of course, it is not possible to demonstrate statistically that a system in a low entropy condition will continuously approach equilibrium thereafter, without making an additional assumption of a stochastic nature, such as the right to throw away correlations at every moment (E), on the justification that they are not macroscopically relevant. But statistical mechanics is precisely a *macroscopic* formalism, in which macroscopic prejudices, such as the irrelevance of microscopic correlations, are legitimate. No one seriously doubts that a complete microscopic description of the system will still show the continuous approach to equilibrium for the vast majority of initial conditions, provided our macroscopic view of 'equilibrium' is retained. Two mixed gases in a flask are no less 'really' mixed even if we are told that if, by some magic, all the molecular velocities were to be reversed simultaneously, the initial unmixed state would be recovered. It is true that this might lead us to call the fully mixed state only a 'pseudo-equilibrium', but this is mere semantics. Throwing away correlations describes *what is actually observed*, and nothing else is needed. If it is accepted that the collective behaviour of atoms, as realized by the motions of macroscopic bodies like pendula, does have significance, then it must be emphasized that microscopic considerations cannot now appreciably affect matters (neglecting now the outside world). It must *not* be imagined that to remove coarse graining removes the macroscopic asymmetry. A complete microscopic description of the pendulum motion would still predict, with overwhelming probability, that the pendulum would be damped to rest, i.e. by retaining the macroscopic concept of 'a pendulum' throughout the microscopic analysis, the same macroscopic asymmetric conclusions are (almost always) arrived at.

A good illustration of the fact that a removal of coarse graining does not remove the macroscopic asymmetry has been given by Rubin[19], using a simple one dimensional model for Brownian motion which may be treated exactly. Rubin considers a lattice of N particles of mass m, elastically coupled to their nearest neighbours, with force constant K, and a larger single particle of mass $M \gg m$. At $t = 0$ the lattice is at rest except for the heavy particle which is given a certain velocity. The subsequent motion of the lattice is then computed. As expected the velocity of the heavy particle is damped, and if an average of its motion is performed this damping is very nearly exponential (for large N) in accordance with 'coarse-grained' expectations. Of course, the overall time symmetry is preserved, and the exact solutions are manifestly invariant under time reversal. Fluctuations occur, and after a length of time of order $\frac{N}{2}\left(\frac{m}{K}\right)^{\frac{1}{2}}$ the system completes a Poincaré cycle, with the heavy particle being 'spontaneously' accelerated to a high velocity once again. If we restrict ourselves (as always with real systems) to lengths of time very much less than the Poincaré time, the asymmetric exponential-damping motion of the heavy particle is assured, for almost all initial conditions, as an *exact* result.

References

1. J. Loschmidt, *Wien. Ber.*, **73** (2), 135, 1876; **73** (2) 366, 1876; **75** (2), 67, 1877
2. E. Zermalo, *Wied. Ann.*, **57,** 485, 1896; **59,** 793, 1896.
3. H. Poincaré, *Acta Math.*, **13,** 1, 1890.
4. K. Huang, *Statistical Mechanics*, Wiley, New York, 1966, p. 85.
5. *The Collected Works of J. W. Gibbs*, vol. 2, Yale University Press, New Haven, 1948, p. 144.
6. P. T. Ehrenfest, *The Conceptual Foundations of the Statistical Approach in Mechanics*, Cornell University Press, Ithaca, 1959, p. 63.
7. R. C. Tolman, *The Principles of Statistical Mechanics*, Oxford University Press, London, 1938, section 51.
8. R. C. Tolman, *The Principles of Statistical Mechanics*, Oxford University Press, London, 1938, p. 176.
9. A detailed account is given in R. Jancel, *Foundations of Classical and Quantum Statistical Mechanics*, Pergamon Press, Oxford, 1969.

10. R. Jancel, *Foundations of Classical and Quantum Statistical Mechanics*, Pergamon Press, Oxford, 1969.
11. E. N. Adams, *Phys. Rev.*, **120,** 675, 1960.
12. K. Huang, *Statistical Mechanics*, Wiley, New York, 1966, p. 90.
13. S. Watanabe, *Rev. Mod. Phys.*, **27,** 179, 1955. T. Wu and D. Rivier, *Helv. Phys. Acta*, **34,** 661, 1961.
14. H. Reichenbach, *The Direction of Time*, University of California Press, Berkeley, 1956.
15. M. Schlick, *Grundzuge der Naturphilosophie*, Vienna, 1948, p. 106. See also A. Grünbaum, *Philosophical Problems of Space and Time*, Knopf, New York, 1963 and Routledge & Kegan Paul, London, 1964.
16. E. Schrödinger, *Proc. Roy. Irish Acad.*, **53,** 189, 1950.
17. See for example J. M. Blatt, *Prog. Theor. Phys.*, **22,** 745, 1959.
18. E. L. Hahn, *Phys. Rev.*, **80,** 580, 1950.
19. R. J. Rubin, *J. Amer. Chem. Soc.*, **90,** 3061, 1968.

Further reading

1. O. Costa de Beauregard, Irreversibility problems, in *Logic, Methodology and Philosophy of Science* (Ed. Y. Bar-Hillel), North Holland, Amsterdam, 1965.
2. R. P. Feynman, The distinction between past and future, in *The Character of Physical Law*, B.B.C. Publications, Cox & Wyman Ltd, London, 1965, chapter 5.
3. R. Jancel, *Foundations of Classical and Quantum Statistical Mechanics*, Pergamon Press, Oxford, 1969. This book also contains an extensive bibliography.
4. P. T. Landsberg, Time in statistical physics and relativity, *Studium Generale*, **23,** 1108, 1970; reprinted in *The Nature of Time* (Eds. J. T. Fraser, F. C. Haber and G. K. Müller), Springer, Berlin, 1972.
5. H. Mehlberg, Physical laws and time's arrow, in *Current Issues in the Philosophy of Science* (Ed. H. Feigl and G. Maxwell), Holt, Rinehart and Winston, New York, 1961.
6. P. Morrison, The instability of the future, in *The Nature of Time* (Ed. T. Gold), Cornell University Press, Ithaca, 1967.
7. R. E. Peierls, Time reversal and the second law of thermodynamics, in *Methods and Problems of Theoretical Physics* in honour of R. E. Peierls (Ed. J. E. Bowcock), North Holland, Amsterdam, 1970.
8. S. Watanabe, Time and the probablistic view of the world, in *The Voices of Time* (Ed. J. T. Fraser), Brazilier, New York, 1966.
9. N. Wiener, *Cybernetics*, Wiley, New York, 1948.

4 Thermodynamics and Cosmology

4.1 The ultimate source of asymmetry

In chapter 2 it was discovered how theromodynamics provided a structural distinction between the two time axis directions. The origin of this asymmetry was traced to the way in which local branch systems, i.e. quasi-isolated regions undergoing entropy changes and accessible to observation and measurement, were generated asymmetrically by interaction with the outside world. It is important to consider now the meaning of the words 'outside world' and what is understood about open systems.

In discussing a laboratory experiment, the outside world could be taken to mean the laboratory itself and all the various external facilities such as heating, lighting etc, as well as the technicians. If, on the other hand, the whole laboratory is to be considered as the quasi-isolated system, the outside world might be taken to refer to the whole earth, or even the solar system. Eventually, of course, the entire universe must be considered. It is possible to identify an ever larger hierarchy of branch systems, each of which comes into being by interaction with a still larger member of the hierarchy.

As an example, recall the discussion of section 3.4 about the formation of a footprint on a sandy beach. The quasi-closed system consisting of the region of the beach in the vicinity of the footprint is formed in a low entropy state by interacting with the 'outside world' which, in this case, is taken to mean the foot of a stroller. At the time of the interaction, when the beach is coupled to the stroller to form a larger system, the local region of the beach undergoes an entropy reduction, but only at the expense of the whole system to which the beach is now open. Specifically, the stroller suffers a metabolic depletion, heat being produced both in his muscles and in the sole of his foot making contact with the ground. Some heat will also dissipate through the air and into the lower layers of sand. The second law of thermodynamics is thereby left intact, but the question of why the total system beach + stroller was not in equilibrium anyway must now be faced, for in order to produce the low entropy footprint in the sand, it is necessary to locate a source of negative entropy in the larger system. But this negative entropy in turn must be accounted for by regarding the stroller as a branch system. This individual is only capable of exercising his locomotive habits as a result of the liberation of the energy of foodstuffs by respiration, using the surrounding atmosphere. His low entropy initial condition is brought about by the interaction with these outside agencies, and the energy dissipated in the interaction brings about a more than compensatory entropy increase. Proceeding further, we are led to photosynthesis by sunlight, and thence to the processes taking place in the interior of the sun, to explain the existence of the consumed foodstuffs in the first place. But even the sun is only a quasi-isolated branch system formed out of the primeval gases at an earlier epoch of the universe.

A search for the origin of the temporal asymmetry of local branch systems leads inexorably backwards and outwards to a discussion of cosmology and the formation of the universe.

4.2 Elementary aspects of modern cosmology

Cosmology is the study of the large scale properties of the universe. In section 1.1 it was mentioned that the geometical properties of space–time were contained in the metric tensor $g_{\mu\nu}(x)$, which is in general a function of position $(x^0 \dots x^3)$, and determines the interval ds through the equation

$$ds^2 = g_{\mu\nu}(x)\, dx^\mu\, dx^\nu. \tag{4.1}$$

Einstein's general theory of relativity provides a connection between the geometrical properties of space–time and the gravitational properties thereof, through the well-known field equations

$$\mathscr{R}_{\mu\nu} - \tfrac{1}{2}g_{\mu\nu}\mathscr{R} = 8\pi G T_{\mu\nu}. \tag{4.2}$$

The so-called cosmic term has been omitted, as is fashionable. (A discussion of cosmological models including this term has been given by Bondi[1].) On the left-hand side of equation (4.2) is a purely geometrical expression. Besides $g_{\mu\nu}$ there is a tensor $\mathscr{R}_{\mu\nu}$ and a scalar $g^{\mu\nu}\mathscr{R}_{\mu\nu} \equiv \mathscr{R}$, formed out of the $g_{\mu\nu}$ and their first and second derivatives (the precise definition need not concern us). On the right-hand side is another tensor $T_{\mu\nu}$, which is the familiar stress–energy–momentum tensor, and is determined by the physical condition of the matter. G is the Newtonian gravitational constant, with appropriate units to connect these two types of quantity.

For a given distribution of mass–energy, the set of equations (4.2) can be solved in principle for the geometrical structure of space in the region concerned. The theory also provides the dynamics, so that the evolution of the geometry with time can be determined. In all but the simplest circumstances, however, the resulting non-linear equations are prohibitively difficult. Nevertheless, it was pointed out in section 1.4 that the smoothed out large-scale structure of the universe could be considered as homogeneous and isotropic, and in uniform motion. This enables the metric to be written in the simple Robertson–Walker form

$$ds^2 = dt^2 - \frac{R^2(t)}{(1 + \frac{1}{4}kr^2)^2}\{dr^2 + r^2\,d\Omega^2\}. \tag{4.3}$$

With this simplified metric, which only contains two unknown scalar parameters, R and k, equation (4.2) may also be simplified to give

$$3\frac{\dot{R}^2}{R^2} + \frac{3k}{R^2} = 8\pi G\varepsilon \tag{4.4}$$

$$2\frac{\ddot{R}}{R} + \frac{\dot{R}^2}{R^2} + \frac{k}{R^2} = -8\pi Gp. \tag{4.5}$$

In these equations p and ε are respectively the pressure and energy density of the cosmological fluid, which is assumed to fill the universe homogeneously. They are the only surviving terms of $T_{\mu\nu}$.

From equations (4.4) and (4.5) the interesting relation is immediately obtained

$$d(\varepsilon R^3) + p\,dR^3 = 0. \tag{4.6}$$

To interpret this result we note that R^3 is a volume which scales with the cosmological expansion. Such a volume is therefore *comoving*. The quantity εR^3 thus represents the energy in a comoving volume, and $p\,dR^3$ is the work done by the pressure in this volume as it expands. Equation (4.6) is just the cosmological analogue of the first law of thermodynamics (conservation of energy) applied to an adiabatically expanding system—compare with equation (2.1).

At our present epoch the ratio $p/\varepsilon \ll 1$, so we may put $p = 0$ in (4.5) and (4.6). Equation (4.6) then reduces to

$$\varepsilon R^3 = \text{constant} = \mathcal{M}, \text{ say,} \tag{4.7}$$

which expresses the fact that the total *rest* mass in a comoving volume is unchanged by the expansion, as expected. If equation (4.4) is rewritten, subject to the condition (4.7) we obtain

$$\dot{R}^2 = \frac{2G\mathcal{M}}{R} - k. \tag{4.8}$$

This equation has three distinct solutions depending on whether $k = -1$,

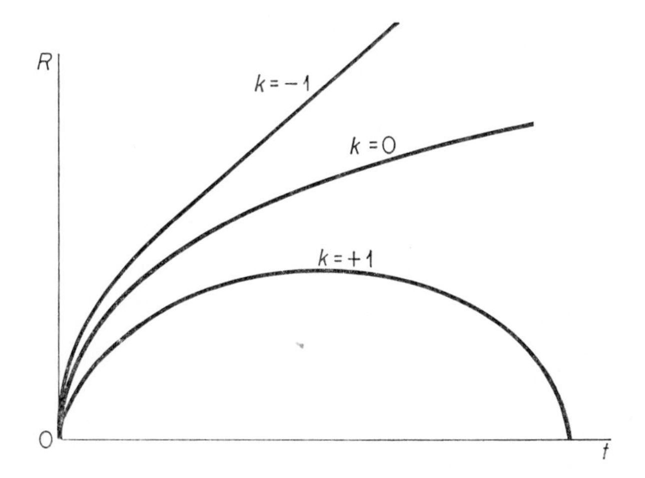

Figure 4.1 The three matter-filled Friedmann cosmological models. All begin at a singular state with $R = 0$ at $t = 0$. The $k = 0$ and $k = -1$ models expand forever, but the $k = +1$ model reaches a maximum value of R, after which it recontracts to another singularity. In a more realistic model regions of small R would be radiation dominated, and expand like $t^{\frac{1}{2}}$.

0 or +1. They are, parametrically,

$$R = G\mathcal{M} \times \begin{cases} 1 - \cos\eta & k = +1 \\ \frac{1}{2}\eta^2 & k = 0 \\ \cosh\eta - 1 & k = -1 \end{cases}$$
$$t = G\mathcal{M} \times \begin{cases} \eta - \sin\eta & k = +1 \\ \frac{1}{6}\eta^3 & k = 0 \\ \sinh\eta - \eta & k = -1. \end{cases} \tag{4.9}$$

The three models described by the solution (4.9) are shown in figure 4.1; they are called the *Friedmann* models after their discoverer[2]. Inspection of equation (4.8) reveals that the k term may be neglected for small R. All three models then expand like $t^{\frac{2}{3}}$. At $t = 0$ the scale factor $R = 0$, and it follows from (4.7) that there is a singular condition of infinite energy density at this point. This event is often referred to as the *creation* of the universe, and the cosmic time t is called the *age*. The $k = 0$ model (also known as the Einstein–de Sitter) continues to expand like $t^{\frac{2}{3}}$, but the models with space curvature $k = \pm 1$ eventually deviate from this as the k term in (4.8) becomes important. The negative curvature $k = -1$ model tends eventually to free expansion, with $R \propto t$, as may be deduced from (4.8) with $R \to \infty$. On the other hand the closed positive curvature $k = +1$ model starts to recontract to small volumes, eventually reaching another singular state with $R = 0$ at $\eta = 2\pi$. This model is symmetric in time about the point of maximum expansion.

It is possible to derive identical behaviour from models based on Newtonian physics rather than general relativity. In this case the behaviour is readily interpreted in terms of the total energy of a comoving volume containing a mass $\mathscr{M}$. This energy is positive when $k < 0$, so that the gravitational attraction of the cosmological fluid is insufficient to prevent complete dispersal of the matter. However, when $k > 0$, the energy is negative, and the fluid remains gravitationally bound, collapsing back again to a dense condition. If $k = 0$ the energy is zero, and the fluid just manages to 'escape' recollapse. The *critical energy density* for this to occur is found from (4.7) and (4.8) to be

$$\frac{3}{8\pi G}\left(\frac{\dot{R}}{R}\right)^2.$$

Among the various simplifications made in deriving these results, one in particular must be questioned and that is the assumption that $p = 0$. Until now nothing has been said about the nature of the cosmological medium. At our own epoch, the major contribution to the energy density ε comes from baryons—stable matter in the form of stars, planets, gas, dust and possibly black holes (see section 4.6). In addition there are various forms of massless particles—photons, neutrinos and gravitons. It is not clear what the energy contribution of the latter two might be, but it is usually held to be insignificant. Photons, mainly in the form of a thermal background (see below), probably provide the dominant contribution, but even this only amounts to 10^{-4} of that of the matter.

On the other hand, matter exerts very little pressure, the dominant contribution being due to the radiation. Now the pressure of radiation is related to the energy density ε_γ by

$$p = \tfrac{1}{3}\varepsilon_\gamma. \tag{4.10}$$

Therefore, as $\varepsilon_\gamma \ll \varepsilon_m$, the energy density of matter, $p \ll \varepsilon_m$ and may be neglected.

However, this will not be true for all R, as can be seen from (4.6). Whereas for the matter

$$\varepsilon_m R^3 = \text{constant}$$

so

$$\varepsilon_m \propto R^{-3}(t) \tag{4.11}$$

for the radiation

$$d(\varepsilon_\gamma R^3) + \tfrac{1}{3}\varepsilon_\gamma\, dR^3 = 0 \tag{4.12}$$

whereupon

$$\varepsilon_\gamma \propto R^{-4}(t). \tag{4.13}$$

Consequently, as $R \rightarrow 0$, then $\frac{\varepsilon_m}{\varepsilon_\gamma} \rightarrow 0$, and the energy density of radiation dominates that of matter.

This is easy to understand physically. The number of particles (either baryons, leptons or photons) in a comoving volume does not change with the expansion (though the photons will be in steady state flow across the boundaries). For non-relativistic matter the energy per comoving volume is simply proportional to the particle mass density, and so remains constant also. This is not true for the relativistic particles, such as photons, for although the photon number remains constant, the photon frequencies (or energies) increase like R^{-1}, rising without limit as the initial state is approached. (This is the well-known cosmological red shift, though in reverse here.)

A new solution of (4.4) and (4.5), with $p = \frac{1}{3}\varepsilon$, must be found for this radiation dominated era. This is readily done: notice that as $R \rightarrow 0$ the curvature term and the $\ddot{R}$ term may be neglected in these equations, which then become identical:

$$R^2\dot{R}^2 = \frac{8\pi G}{3}\varepsilon R^4 = \text{constant},$$

so $R \propto t^{\frac{1}{2}}$.

From this result it appears that whichever model is chosen a condition of infinite density, with $R = 0$, cannot be avoided at $t = 0$. It is not necessary to dwell on the subtleties of this, except to mention that the condition cannot be avoided by minor departures from homogeneity and isotropy. For our purposes it need only be accepted that the universe is supposed to have been in a state of very high compression in its early stages, according to these models. Some evidence that this was indeed the case will be discussed below.

Just how long ago this singularity may have occurred depends on the constants $\mathscr{M}$ and k in (4.8), which in turn depend on the density of matter and the geometry, both of which are in principle accessible to observation and measurement. In practice, measurement of the curvature, from the angular sizes of galaxies, cannot be carried out with any sort of reliability. On the other hand, the density of luminous material is quite well known, working out at about 10^{-31} g cm^{-3}, which is about 1 % of the critical density. The only problem is that we can never be sure that substantial quantities of matter do not exist which are escaping detection. This would be so if there were large amounts of intergalactic gas, or black holes (see section 4.6).

An independent check on these results is provided by the direct measurement of $\dot{R}/R$, and the higher derivative $q = -R\ddot{R}/\dot{R}^2$, known as the deceleration parameter. Current values for these quantities[3] are about 10^{-18} sec^{-1} and 1 $\pm$ 0·5 respectively, which gives an 'age' for the universe (i.e. time since the singularity as measured by a comoving observer) of about 10^{10} years, depending slightly on the model. This age is consistent with the estimated age of the Earth as deduced from the lifetimes of some radioactive substances. The value of $q = +1$ should not be taken too seriously, because of the large uncertainty. At face value it indicates a closed model of the cyclic kind, but as this requires ε to be greater than the critical density, it is in apparent contradiction with the density observations.

The enormous compression of the early stages generates very high temperatures, a condition in which matter becomes strongly ionized and coupled to the radiation, which is thereby thermalized. This thermal radiation will cool as the universe expands, but will always be present as a black body background isotropic in the comoving frame. The fact that such radiation has actually been observed[4], at a temperature of 2·7 K, lends remarkable support to the hypothesis of an early dense phase of the universe. Using the value of this temperature, which amounts to a photon/baryon ratio as high as 10^9, it may be calculated that the universe was radiation dominated until 10^3 years after the singularity. The entropy density of this radiation is proportional to the photon density, which varies like R^{-3}: hence the entropy per baryon is constant under the expansion (neglecting interaction). Because this ratio is so high the universe must really be considered as very hot. All the entropy producing processes that surround us, including those vital to the existence of human life, contribute only a small fraction to the entropy content present since the beginning of the world.

4.3 The Hubble law and horizons

It was briefly mentioned in section 1.4 that the universe appears to be expanding uniformly in our local frame. This imples that the large scale condition of the world can be labelled by a single function of the cosmic time t. This function is the scale factor $R(t)$ for the distance between comoving observers. A helpful way of visualizing this expansion is to imagine the three-spaces at constant t as represented by a 2-surface rubber sheet being stretched uniformly in all directions (see figure 4.2). The galactic clusters are imagined as dots on the sheet, and a few moments thought reveals that any given dot appears to be moving away from all its neighbours—every such dot would see itself at the 'centre' of an expanding system of dots.

The rubber sheet analogy can even be used to represent models with curvature. For example, a spherical shell (a balloon) being inflated and deflated corresponds to the oscillating $k = +1$ model, with positive curvature and finite volume.

Now if the expansion is everywhere uniform, dots which are a certain

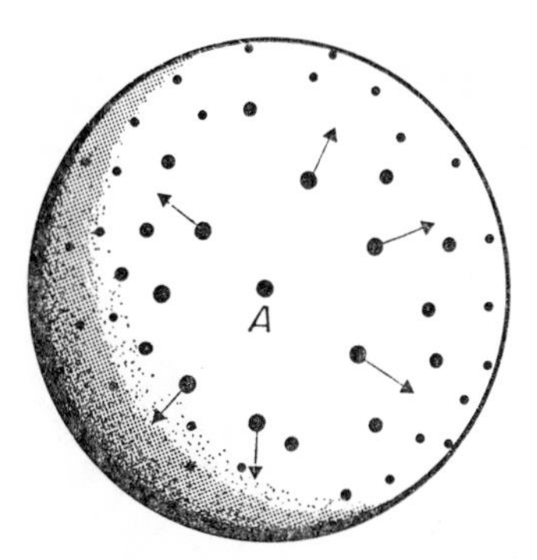

Figure 4.2 The expanding balloon analogy. As the membrane stretches every dot moves away from every other. Any given dot, such as A, will consider itself at the centre of the expansion, even though the dots are in fact always distributed homogeneously. The singular condition of the universe corresponds to complete deflation of the 'ballon' to a point.

distance apart will recede from one another twice as fast as dots separated by one half that distance. If we could view the whole of our universe in a comoving frame instantaneously, we should deduce the law that the speed of recession v, of a distant galaxy, is proportional to its distance r:

$$v = Hr. \tag{4.14}$$

H is a constant of proportionality with dimensions time^{-1}; it is just the quantity $\dot{R}/R$.

In practice it is not possible to view the whole three-space of our world simultaneously. What is actually seen is a set of points on the backward null cone from the point of observation. Therefore, although the expansion might be everywhere uniform, the delay caused by the travel time of light from distant matter might be so long that the expansion rate H may have changed in the meantime. (Remember R is a function of t, so that H will be time dependent unless $R \propto e^{Ht}$). Nevertheless, on a small scale equation (4.14) will be approximately true, and this relation was first confirmed by Hubble in 1929 from measurements of the red shifts in the spectra of distant galaxies[5]. H is therefore known as the *Hubble constant*, and H^{-1}, which is of the order of the age of the universe, is called the Hubble time.

The consequences of the cosmological expansion for light propagation are interesting. In figure 4.3(a) the wavy arrow represents a photon emitted by a distant galaxy B in the direction of A, which we may take to be our own galaxy. Because of the expansion, A and B are receding, so it may happen that the photon will never reach us. If that were so, there would be events in the universe which are forever unobservable by us. These events are separated from the observable ones by the spherical surface of photons that converge onto

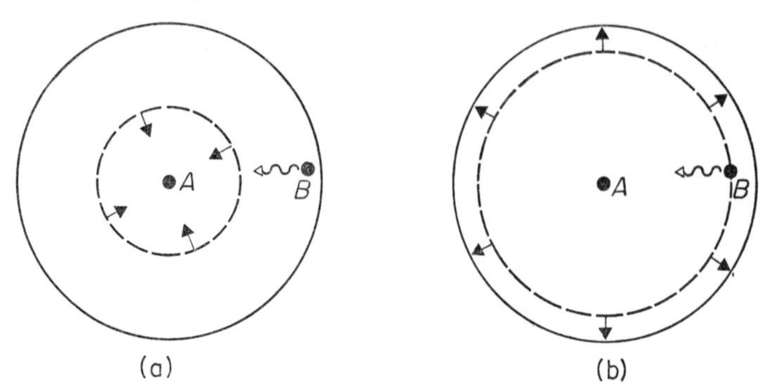

Figure 4.3 (a) Event horizon. The broken circle represents a wavefront of light which just reaches A at $t = \infty$. The photon from B will never be observed. (b) Particle horizon. The broken circle represents the wavefront of light emitted by A at the creation event. An observer at A will be able to observe B for the first time when this wavefront sweeps across it.

us at $t = \infty$. This surface is therefore referred to as an *event horizon*, and may be thought of loosely as the region at which the galaxies are receding from us at the speed of light. The steady-state model, for which the expansion rate $\dot{R}$ is always accelerating ($R \propto e^{Ht}$) possesses an event horizon.

If the photon from B was emitted at the creation event at $t = 0$, it may be that the expansion is such that it has not yet reached us, but will do so eventually. When it does, we should be able to see the galaxy B for the first time, and conversely B would be able to observe us for the first time. It follows that the wavefront of light emitted by our galaxy at the creation event (see figure 4.3(b)) divides the other galaxies into two classes: those inside the wavefront, which can be seen already, and those outside which cannot yet be seen because of the finite age of the universe. The Friedmann models described in section 4.2 all possess this type of horizon, called a *particle horizon*, though they do not possess event horizons.

Events which are separated by horizons cannot causally influence one another. They are important in the discussion of density perturbations in the early universe, and in the theory of black holes.

4.4 Thermodynamics of simplified models. Reversible cycles

In passing from the Minkowski to the Robertson-Walker metric space, the subject of cosmology introduces a number of novel and important features into the subject of thermodynamics. The Friedmann models described in section 4.2 bear a certain resemblance to the adiabatic expansion of a gas inside a piston and cylinder arrangement. Of particular interest is the closed

Friedmann model, which expands to a maximum value of R, and then re-contracts in a cyclic fashion, similar to the raising and lowering of a piston. However, a certain amount of caution is necessary in drawing cosmological conclusions from laboratory arrangements of this type. Some of the important differences are illustrated in two highly idealized models due to Tolman[6].

In the first model, suppose that the universe is uniformly filled with black body electromagnetic radiation, or any extreme relativistic fluid. The wave-lengths λ scale with the expansion factor $R(t)$, so that the radiation frequency ω is a function of t:

$$\omega \propto R^{-1}(t) \qquad (4.15)$$

which is the basis of the red shift discussed in section 4.3. Similarly, the radiation temperature T_γ, which is proportional to the average frequency, has the dependence

$$T_\gamma \propto R^{-1}(t). \qquad (4.16)$$

As all wavelengths scale in the same ratio, the expansion will not affect he distribution of energy among the frequencies (the spectrum shape). The entropy density of black body radiation is given from elementary thermo-dynamics by the formula $\frac{4}{3}aT_\gamma^3$, where a is a constant with the dimensions of erg cm^{-3} deg^{-4} and is known as Stefan's constant; it has the value $7{\cdot}6 \times$ 0^{-15}. It then follows from (4.16) that the entropy in a comoving volume is onstant. (In relativity theory, entropy must be redefined in a covariant ashion. However, for a comoving observer performing experiments only in his immediate neighbourhood, this reduces to the usual definition—see Tolman[7].) In other words, the expansion of the universe does not change the entropy of an extreme relativistic fluid. This is in fact the same as the labora-tory situation in which an arbitrary expansion of a radiation filled cavity does not change the entropy. However, the reader should be careful not to confuse the motion of the expansion with the motion of the radiation. In the real universe there are no comoving *bounded* volumes in which we may magine the radiation to be confined. Instead, we may imagine an invisible transparent 2-surface delineating the volume, with photons continually rossing to escape from the inside. However, if the space is homogeneously filled with radiation, photons will enter the volume from the outside at the ame rate, so the average number of photons in any such volume is constant. ndeed, if the photon density is called ρ_γ this must be equal to $\varepsilon_\gamma\omega^{-1}$, which, rom (4.13) and (4.15), is proportional to R^{-3}. Hence

$$\rho_\gamma R^3 = \text{constant}. \qquad (4.17)$$

The second of Tolman's models is a universe uniformly filled with non-relativistic particles; for example, a monatomic ideal gas. Once again, the xpansion cannot change the velocity distribution of the gas, because a re-scaling of lengths merely rescales the velocities in the same ratio. For a non-relativisitic gas the velocity and momentum distributions coincide, so i the gas is in thermal equilibrium with a Maxwellian distribution, it wil remain in equilibrium, and the entropy will not change. This is easily demon strated. The de Broglie wavelength λ of an individual particle scales witl $R(t)$, so the particle momentum, $2\pi\lambda^{-1}$ in our units, varies like R^{-1}. Henc the kinetic energy varies like R^{-2}. The temperature of the gas, T_m, is pro portional to the root mean square kinetic energy, and so must also vary lik R^{-2}:

$$T_m \propto R^{-2}(t) \qquad (4.18$$

a result which should be compared with (4.16). Now, the entropy of a mon atomic ideal gas in a volume R^3 is known from elementary thermodynamic to be proportional to log $(R^3T_m^{3/2})$, which is constant because of (4.18).

The reader should not imagine that (4.18) is a purely quantum mechanica result. In an expanding universe, an arbitrarily moving classical particle wil always appear to lose energy with respect to the local comoving frame. If w imagine such a particle with velocity v in this privileged frame, after a time d it will have travelled a distance $v\,dt$ to a new position where the local privilege frame is in motion with respect to the first because of the expansion. Thi motion has a velocity $\frac{\dot{R}}{R}(vdt)$ away from the original position, irrespective o the direction (equation (4.14)). Consequently, the velocity of a particl relative to a comoving observer at the new position will have been reduced b an amount $dv = -\frac{\dot{R}}{R}(vdt)$. Integrating this equation leads to the resul $v \propto R^{-1}(t)$, and hence (4.18).

It is important to note that equation (4.18), and the conservation of entrop which follows as a consequence, are independent of the collisions betwee the molecules. There is no specified finite relaxation time (collision time) fo the gas against which the expansion time scale can be compared. The con clusions are therefore not restricted to quasi-static expansions. Actually equation (4.18) may be recovered easily from ordinary thermodynamics. Th temperature of an ideal gas is related to its volume R^3 along an adiabatic patl by the equation $T_m(R^3)^{\alpha-1}$ = constant, where α is the ratio of the principa specific heats. For a monatomic gas $\alpha = \frac{5}{3}$, and (4.18) follows immediately

These results will now be examined from the standpoint of reversibility In both the non-relativistic and extreme relativistic fluid-filled models, th total entropy in a comoving volume is constant. In the recontracting Fried mann model, $R(t)$ eventually returns to its particular value with the sam temperature and entropy density, so we should be inclined to call thes idealized models macroscopically *reversible*. Actually, any observer woul see during the expanding phase an apparently irreversible flow of energ

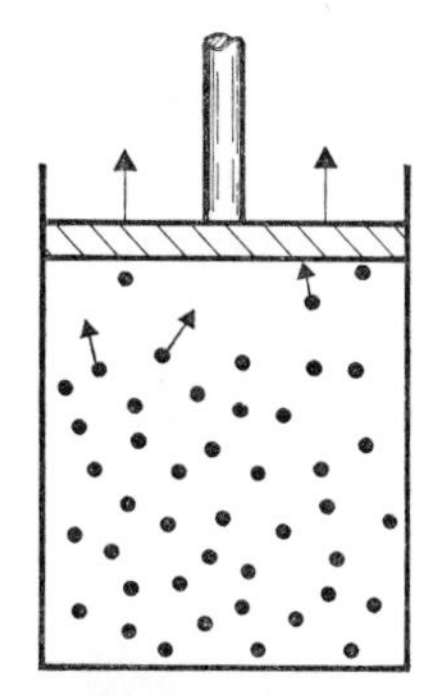

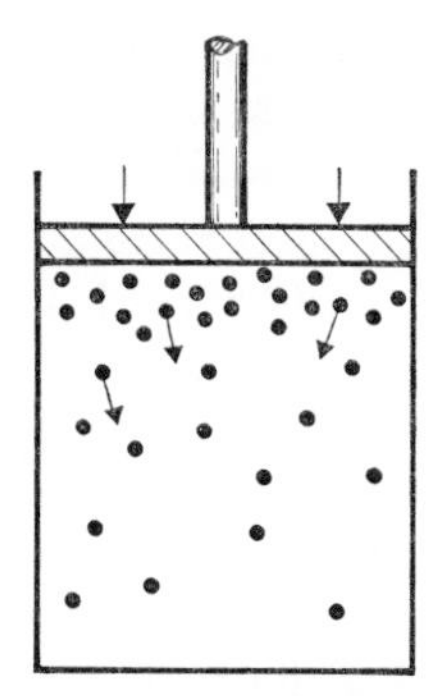

Figure 4.4 Expansion and contraction of a gas by a piston at a finite rate. The gas molecules will always lag a little behind the piston motion, thereby destroying the equilibrium and increasing the entropy. In the cosmological case the expansion occurs uniformly throughout, so that equilibrium is maintained and the entropy remains constant.

away from his local vicinity into the colder depths of space. However, a subsequent recontraction would merely reverse the flow and restore the situation.

In laboratory thermodynamics, an adiabatic expansion of non-relativistic matter can only be isentropic if it is also quasi-static (carried out infinitely slowly). In the cosmological case the expansion occurs at a finite rate, which may appear rather strange at first sight. The causes of entropy increase in an adiabatic piston and cylinder expansion may now be considered[8] (see figure 4.4). Firstly, there is friction between the moving parts of the container. Secondly, a rapidly moving piston will set up turbulence and temperature gradients which, on restoration of equilibrium, increase the entropy. Finally, the finite relaxation time of the gas implies that the pressure behind the retreating piston as it is withdrawn is less than elsewhere, because the molecules take a finite time to rush into the opening space. More work is done in recontracting the gas (the pressure beneath the piston is now greater than elsewhere) than is gained from its expansion. The first law of thermodynamics tells us that this extra work appears as heat, and hence entropy.

During a cosmological expansion and contraction none of these effects operates, because the motion is always taking place under equilibrium conditions. There is obviously no moving container, nor can the expansion produce temperature and pressure gradients, because this would contradict the assumed homogeneity of the fluid. The expansion occurs uniformly throughout the cosmological medium.

It is clearly an oversimplification to treat the contents of the universe as a homogeneous single-component fluid. Slightly more realistic would be a universe homogeneously filled with non-relativistic matter and electromagnetic radiation together. In the absence of interaction between them, a temperature difference will develop as the universe expands, because the matter cools faster than the radiation—see equations (4.16) and (4.18). If the coupling is 'switched on' energy will tend to flow from the radiation field to the matter in an attempt to equalize the temperatures (whether or not this equalization is achieved will be discussed in section 7.1). If the expansion time scale is shorter than the temperature equalization time, this heat flow will take place under non-equilibrium conditions and the entropy will be increased. Moreover, if the expansion is followed by a recontraction, eventually there will be a flow in the opposite direction, also accompanied by an entropy increase.

There are other entropy increasing processes present, even in very simple fluids. For small values of R the matter will be partially relativistic and the expansion will modify the equilibrium energy distribution, so that the entropy increases as the gas relaxes back. For a realistic model gas, e.g. hydrogen, there will be other processes, such as ionic recombination and molecular formation, which will increase the gas entropy. In some cases the relaxation times for these processes may be larger than the expansion time scale.

4.5 Entropy of starlight. Olbers' paradox

Although the increase in entropy density arising from the type of processes discussed in the previous section might be quite large numerically, their relevance to the kind of irreversible phenomena of significance to terrestrial observers is questionable[9]. By far the most significant source of entropy in our environment is the steady production of electromagnetic radiation by the sun. Not only is this quantitatively the most copious source, it is qualitatively the most important, for the very existence of life on Earth depends on the thermodynamic disequilibrium which this energy flow produces.

It is not difficult to appreciate that the sun is a source of entropy increase, for it is clearly in a non-equilibrium condition with its environment. The central temperature is some tens of millions of degrees, but the surface temperature is only a few thousand degrees, while the temperature of surrounding space is as low as three degrees (see section 4.2). It follows that a continual transfer of energy away from the centre, towards the surface and into space, is taking place along an enormous temperature gradient. This energy flow is maintained at the expense of the rest mass of the core nuclei which are undergoing nucleosynthesis. Specifically, hydrogen 'burning' provides the greatest energy, and occurs in the early stable period of the lifetime of a star. The high core temperature enables the protons to overcome the Coulomb potential barrier, and, after a series of nuclear reactions, this results in fusion to form helium. For example, the following series is known as the proton–proton chain:

$$\mathrm{p(p,\ \beta)D}$$

$$\mathrm{D(p,\ \gamma)He^3}$$

$$\mathrm{He^3(He^3,\ 2p)He^4}.$$

The result of these reactions is to liberate about 1% of the rest mass of the 'cooked' protons in the form of γ rays and neutrinos. The neutrinos immediately leave the interior of the sun and pass off into space without appreciably interacting with the outer layers, which are transparent to neutrinos. They do not represent a significant increase in entropy. (This is also true by virtue of the fact that neutrinos carry a conserved quantum number and so cannot be irreversibly degraded to a greater number of lower energy neutrinos, as is the case with photons, and gravitons.) On the other hand, the γ rays are strongly coupled to the dense, highly ionized particles surrounding them. In particular, their mean free path against electron Compton scattering is very small, and their energy takes millions of years to get out to the surface. During this time there will have occurred a tremendous number of interactions, each having the effect of dispersing the energy among more and more degrees of freedom. Thus, when this electromagnetic energy departs from the surface of the sun, it is no longer in the form of MeV γ rays, but of photons in the visible region, some 10^7 times less energetic. It follows that every proton cooked in the sun (or any other similar star) creates about 10^7 starlight photons, each photon representing a new degree of freedom; this amounts to an entropy increase of about 10^7k. There will also be a further increase as these starlight photons eventually interact with galactic and intergalactic matter. However, the mean free path in intergalactic space is about a Hubble radius (10^{28} cm), so this will happen very slowly; it is, of course, the same as the process discussed in the previous section. Galactic absorption is more important in the short term, and the photons stopped by the cooler Earth cause, directly or indirectly, the increase in entropy so familiar in everyday life—the weather processes, snow melting, erosion, biological growth, etc.

The thermodynamic properties of starlight are intimately connected with a well-known paradox due to Olbers[10]. In its original form it was applied to a static Euclidean universe of infinite extent and uniform density of luminous matter. If we lived in such a universe, every line of sight would eventually intersect a star; so that each point in the sky would appear as bright as the sun's surface. The argument is unaffected by the distances at which the stars lie, because although the total luminosity of an object is inversely proportional to the square of the distance from the observer, so is the surface area; the energy per unit area of emitting surface is therefore independent of distance. The conclusion is that the simple model universe described is inconsistent with the darkness of the night sky.

The same conclusion may be reached by considering the flux from successive concentric shells of stars surrounding the observer. For a uniform source density, the number of sources in a shell of radius r and thickness dr is proportional to $r^2\,dr$, while the flux from a given source is proportional to r^{-2}. (These arguments are unaffected by considerations of the curvature of space, which enters both constants of proportionality in a compensatory way.) It follows that the contribution from all shells is the same, whatever the distance. In an infinite universe the flux apparently diverges. Actually this is not quite correct because reabsorption of the radiation by the stars would occur (in the previous argument this corresponds to their surface areas overlapping). Instead, the stars would be in equilibrium with space at a very high temperature.

It is easy to see that the 'paradox' arises as a consequence of the static, infinite, uniform nature of the cosmological model employed. A resolution is immediate in any model which is either of finite age, or expands sufficiently fast (or both). This is best understood by returning to equation (4.12), which describes the adiabatic expansion of source free radiation in a comoving volume R^3. The effect of sources can be taken into account by adding to the right-hand side the rate of energy emission in the form of starlight from sources with density ρ and luminosity L

$$\frac{d}{dt}(\varepsilon_\gamma R^3) + \tfrac{1}{3}\varepsilon_\gamma \frac{dR^3}{dt} = L\rho. \tag{4.19}$$

This equation may now be integrated over the past to give the total radiation energy density at the (present) time t

$$\varepsilon_\gamma(t) = R^{-4}(t)\int^t L(\tau)\rho(\tau)R^4(\tau)\,d\tau. \tag{4.20}$$

If the number of stars (or galaxies) is considered as fixed, ρR^3 is constant and may be removed from the integral

$$\varepsilon_\gamma(t) \propto R^{-4}(t)\int^t L(\tau)R(\tau)\,d\tau. \tag{4.21}$$

The expression on the right-hand side of (4.21) will always converge if R increases fast enough. For instance, in a universe with sources of constant luminosity, L, this will be so if $R \propto t^{\frac{1}{3}}$ or faster, even if the lower limit on the integral is $-\infty$. Physically, the divergence is avoided because the distant galaxies are red shifted by the expansion, so that their contribution to the energy density at a given point is progressively diminished with their distance from that point. Alternatively, if the lower limit to the integral is finite, the expression on the right of (4.21) will always converge, even for a static universe $R =$ constant (e.g. Olbers' model). In this case the divergence is avoided because a given observer would not be able to see beyond a sphere whose radius is equal to the light travel distance over the age of the universe.

It has been mentioned already that real sources have a finite surface area, so that if the observer's line of sight is extended far enough the sources will start to overlap. Thus, sources beyond a certain distance, say r_0, will be obscured by the nearer sources, i.e. an observer could not see beyond a distance r_0 anyway, so there will be no divergence. Nevertheless, if the light travel distance over the age of the universe were equal to r_0 the night sky would still appear as bright as the surface of a star, in contradiction with observation. To estimate r_0 it is supposed that the average area of a star is A. The number of sources in a shell of radius r is $4\pi r^2 \rho \, dr$, so that the fraction of the shell area covered by sources is $A\rho \, dr$. Superimposing all shells, the total solid angle of the sky covered by luminous matter will be 4π when this fraction is unity:

$$A\rho \int_0^{r_0} dr = 1$$

i.e.

$$A\rho r^2 = 1. \tag{4.22}$$

The time required for light to travel the distance r_0 is therefore $(\rho A)^{-1}$ in units with $c = 1$. We conclude that a static, Euclidean, uniform universe is only consistent with the darkness of the night sky if its age is $\ll (\rho A)^{-1}$. After times $\simeq (\rho A)^{-1}$ the stars act as an adiabactic enclosure, with the temperature of space reaching that of the stellar surfaces. This is the condition for thermal equilibrium, so that $(\rho A)^{-1}$ measures the *relaxation time* of the universe for starlight emission.

This result is perhaps better understood from the general equation (4.19). If two stars overlap, the more distant one is not seen because its light is absorbed by the nearer one. Therefore, a term ought to be included on the right-hand side, corresponding to absorption of starlight; this term would remove any divergence from (4.21). The upper limit on this quantity would then be determined by the equilibrium situation in which the rate of emission of radiation was equal to the rate of absorption. For a given star of constant luminosity the energy emission rate is L, so that the energy density of starlight is $L\rho t$ after a time t. For stars of average surface area A the rate of absorption will be $L\rho tA$, with equilibrium occurring when this quantity is equal to the emission rate L. This occurs after a time $t \simeq (\rho A)^{-1}$, which is the result already found.

Let us now summarize this result. In a static universe of finite age filled homogeneously with constant luminosity stars, the stars act as localized heat sources in a cold space. Gradually, as the starlight accumulates, the temperature of space rises, until eventually equilibrium is reached when every star is absorbing energy from the surrounding space at the same rate as it emits (this is the situation Olbers wished to avoid). As this condition is approached, the entropy of the system assumes its constant maximum value. On the other hand, if the universe expands faster than $t^{\frac{1}{3}}$, then however long the stars burn this equilibrium will never be reached, and there will always be a temperature gradient around the stars.

An estimate for the relaxation time for this process will now be given. In the present state of the universe there are about 10^{20} stars in the $(10^{28})^3$ cm^3 volume of space inside the particle horizon. The average density of stars is approximately 10^{-64} cm^{-3}. Taking the sun as a typical star, the surface area is of order 10^{23} cm^2, giving a value for the relaxation time of order 10^{30} sec or 10^{23} years. The most significant feature about this result is that it is enormously longer than any cosmological time scale (the Hubble time is about 10^{10} years). Consequently, provided only that a disequilibrium exists, the expansion of the universe is irrelevant to the emission of starlight.

The relation between the expansion of the universe and the flow of starlight appears to have been decisively misunderstood by some authors[11]. In section 4.4 it was demonstrated that a universe homogeneously filled with radiation could not change its entropy by expansion. What appears to a local observer to be an irreversible flow of radiation out of his region of space, is really reversible; a subsequent recontraction merely restores the same temperature and photon density as the radiation flows back again. This is in sharp contrast to the emission of starlight, which also appears to a local observer near a star to be an irreversible flow of radiation away from his region of space. The latter process is, however, genuinely irreversible, because photons are made, whereas in the former process they are conserved. Starlight emission generates entropy because it takes place in disequilibrium associated with local departures from homogeneity.

The irreversible nature of starlight is well illustrated by considering the outcome of a subsequent recontraction of the universe, such as occurs in the closed Friedmann model. As remarked above, changes in R do not affect the emission of starlight, and the stars will continue to emit radiation in the same way after the contraction begins. Of course, the energy density of radiation in space will begin to rise gradually due to the contracting volume, but it will not reach the temperature of the stellar surfaces until very near the final dense state. Rees[12] has shown that the stars would actually be destroyed by collisions before they were vaporized by the radiation. The contracting phase does not therefore restore the changes made in the expanding phase; things continue in the same way. This would be well appreciated by a local observer, who would find nothing remarkable about the onset of contraction. The distant galaxies would remain red shifted, because of the time taken by the light to travel the intervening distance; only the nearby galaxies would start to become blue shifted. This observer would guess nothing of the contraction of the universe until many millions of years later, when the more distant galaxies became blue shifted also. This irreversibility of starlight, and other processes, destroys the exactly reversible nature of the closed Friedmann models discussed earlier (see also sections 7.3 and 7.4).

4.6 Entropy and gravitation

The relevance of gravitation to the subject of entropy increase is twofold. Firstly, the growth of density perturbations by gravitational concentration of the cosmological fluid is responsible for the existence of the inhomogeneous structures such as stars, necessary for the production of entropy by starlight. Secondly, gravitational condensation and collapse is itself a means of bringing about an increase of entropy.

The application of thermodynamics to a self-gravitating system is a highly non-trivial matter, concerning which only limited progress has been made[13]. The long range nature of gravitational forces introduces novel features into the equilibrium properties of the system. A non-gravitating system subject to only short range forces may be maintained in equilibrium by appropriate control over boundary conditions. The usual equilibrium configuration of a gas is one of uniform temperature and density. On the other hand, a self-gravitating system can only be constrained to remain in equilibrium by *internal* forces, so that these systems do not 'naturally' reside in a uniform state, but tend instead to collapse. In other words, the natural fluctuations are much larger than the usual '$N^{\frac{1}{2}}$' for non-gravitating systems. Formally, the entropy of a gravitating system has no maximum, so that there appears to be no equilibrium state at all; complete collapse is permitted. In practice, internal forces usually prevent this fate, allowing local entropy maxima (see figure 4.5). These metastable states are of several varieties; for example, stars are supported by kinetic or electron degeneracy pressure, neutron stars by neutron degeneracy pressure, etc.

An interesting model of a self-gravitating fluid has been given by Saslaw[14], based on an analogy with the van der Waals equation of state. There appears to be a close similarity between a phase transition in which a vapour condenses to a liquid, and the condensation of discrete objects from a gravitating fluid. In this model a system will tend to maximize its entropy by undergoing collapse—a condition of uniform density is not an equilibrium state. An initially smooth fluid will try to fluctuate wildly, and grow structure; as with the growth of a crystal, the highest entropy state is a structured state. Of course, the appearance of this structure (which represents information, or

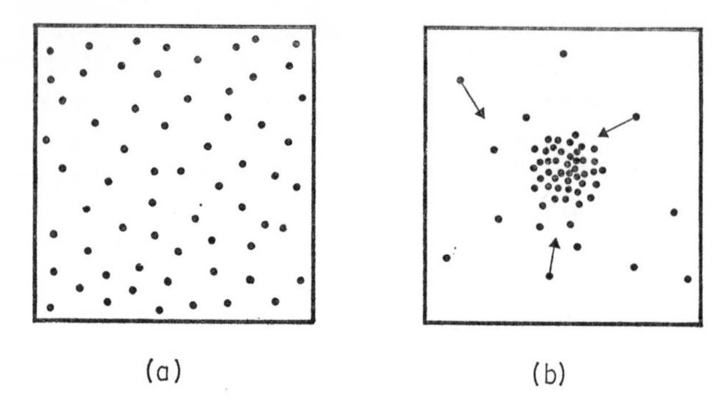

Figure 4.5 In (a) a confined gas is depicted in which gravitational effects have been 'switched off'; it is in equilibrium at uniform density. In (b) there is appreciable self-gravitation. The particles collapse to a new configuration, increasing the entropy of the system. The new configuration is only a pseudo-equilibrium though, for if the gas were squeezed sufficiently it would collapse without limit to a black hole—a state of infinite density and entropy.

negative entropy) is compensated by a corresponding entropy increase in the form of internal heating and radiative processes.

Any truly satisfactory theory of gravitational thermodynamics must include the effects of general relativity. It has already been mentioned that real systems are observed to be in local metastable states because internal pressure effects support them against collapse. However, according to Einstein's theory, if a given mass is compressed to a sufficient density it will completely collapse, irrespective of the nature of the internal pressure[15]. This is because, in general relativity, pressure is also a source of gravitation; for an object inside a critical radius, known as its Schwarzschild radius, the gravitating effect of pressure overwhelms its supportive effect, and the whole mass implodes to a zero volume and infinite density. This total collapse to a singular state occurs in a very short time as measured by an observer falling in with the collapse, though this time is infinite for an asymptotic observer.

In realistic situations such an object becomes surrounded by an event horizon (see also section 4.3). No energy of information may then pass out of the object to the surrounding universe, though anything may pass inwards. It appears that the only parameters that can be used to describe the internal state from the outside are the total mass, charge and singular momentum; all other physical quantities become unmeasurable[16]. Because no light can escape from such an object after a very short time, it is called a *black hole*. It is widely believed that black holes will form as the end point of evolution of stars a little bit more massive than the sun. If this is so, they will be a common type of object in the universe. At the time of writing there is no conclusive observational evidence to support this speculation.

What can be said about the entropy of a black hole? As regards the interior, it appears that this entropy increases without limit during the collapse, as more and more gravitational energy is pumped into the internal degrees of freedom. However, for an outside observer, the black hole only has three allowed degrees of freedom (mass, charge, angular momentum) and the measurement of internal entropy is impossible. Thermodynamically, the object behaves like a black body at zero temperature with an infinite thermal capacity; an unlimited amount of work may be put into it without raising the temperature. Thus, whenever entropy disappears down a black hole, the second law of thermodynamics appears to be violated[17].

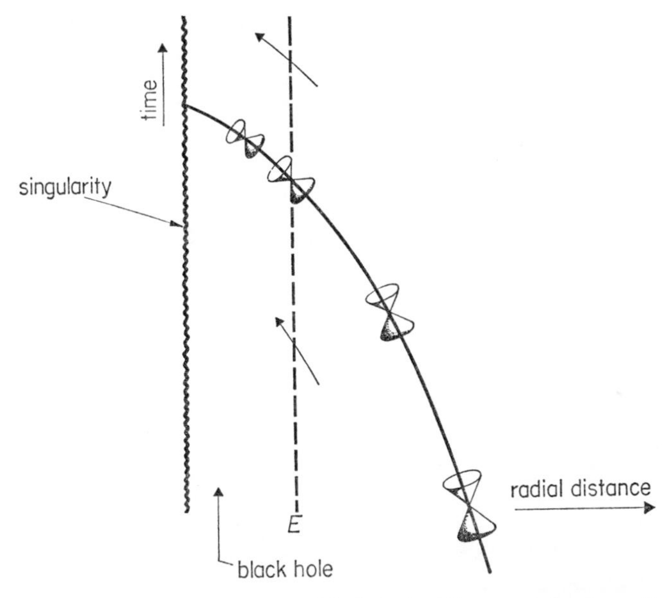

Figure 4.6 Time asymmetry near a black hole. The event horizon E acts as a one-way membrane; particles can pass through into the black hole, but not the other way. This happens because the space around the object is distorted so that the null cones become more and more tilted as E is approached. Inside E the tilting is so great that all timelike world lines must end on the singularity. No amount of energy can bend them back through E.

In compensation, black holes display their own distinctive kind of irreversibility. The very existence of an event horizon is one example; it has the property of a one-way membrane allowing only the passage of energy from the outside to the inside, but not the other way (see figure 4.6). In the case of rotating black holes, an interesting irreversible process has been discussed by Penrose and Christodoulou[18]. The squared mass M^2 of such an object may be decomposed into two parts (units with $G = c = 1$ are used here)

$$M^2 = M_{ir}^2 + \frac{L^2}{4M_{ir}^2}. \tag{4.23}$$

L is the angular momentum, and M_{ir} is known as the irreducible mass for the following reason. It is possible to devise processes in which the rotational energy of the black hole is extracted by the outside; this energy is associated with the second term in (4.23). However, unless these processes are of a very special character ('reversible'), they inevitably lead to an irreversible increase of M_{ir}. It follows that M_{ir} can never decrease (the maximum amount of energy that can be removed in this way is in fact 29%). Another way of expressing this result is to say that the event horizon area, which turns out to be $16\pi M_{ir}^2$ can never decrease. M_{ir}^2 therefore plays a role in black hole physics very similar to the entropy in thermodynamics. Indeed, if equation (4.23) is differentiated, the following is obtained:

$$dM = \Pi\, d\sigma + \mathbf{\Lambda} \, . \, d\mathbf{L} \tag{4.24}$$

with

$$\sigma = M_{ir}^2, \quad \Pi = \frac{1}{2M}\left[1 - \frac{L^2}{4\sigma^2}\right] \quad \text{and} \quad \mathbf{\Lambda} = \mathbf{L}/(4M\sigma).$$

Equation (4.24) should be compared with the first law of thermodynamics (equation (2.2))

$$dE = T\, dS - p\, dV \tag{4.25}$$

so identifying S with σ and the external work $p\, dV$ with $-\mathbf{\Lambda} \, . \, d\mathbf{L}$, the extracted angular momentum energy. The analogue of the second law is

$$d\sigma \geqslant 0, \tag{4.26}$$

a result known as Hawking's theorem[19]. It expresses the fact that any changes occurring to black holes must increase their area (including the merging of two such objects). There is even an analogue of the third law of thermodynamics.

Bekenstein[20] has proposed a new 'generalized' second law aimed at avoiding the violation of the ordinary thermodynamic second law when entropy falls into a black hole. Making use of the irreversibility of Hawking's theorem, he proposes that the *generalized* entropy

$$S_T = S_{ext} + S_{bh} \tag{4.27}$$

be defined, where S_{ext} is the 'usual' entropy of the external world, and S_{bh} is the black hole 'entropy' σ. The generalized second law then becomes

$$dS_T \geqslant 0 \tag{4.28}$$

which embodies the usual second law and Hawking's theorem as special cases. Bekenstein even suggests a quantum mechanical expression for S_{bh}:

$$S_{bh} \propto \frac{k\sigma}{l^2} \tag{4.29}$$

where k is Boltzmann's constant and l is the Planck length $\left(\frac{\hbar G}{c^3}\right)^{\frac{1}{2}}$; the constant of proportionality is suggested to be about 96 or less. Notice that S_{bh} is finite in quantum mechanics (related to its finite size as a quantum mechanical 'box'), but tends to infinity as $\hbar \to 0$.

Finally, a few words must be said about the relationship of the generalized entropy to statistical mechanics. In thermodynamics we are restricted to measurement of a few macroscopic parameters, such as temperature and pressure, although there may be many microscopic configurations that realize a given macrostate. Similarly, in black hole physics, an external observer is restricted to measurement of only three parameters, M, L and the electric charge, despite the fact that internally the black hole has many degrees of freedom. As discussed in chapter 3, it may happen that a rare fluctuation in a thermodynamic system will bring about a decrease in the system entropy S_{ext}. It may therefore be wondered whether a comparable fluctuation in black hole entropy S_{bh} is possible, only, of course, after the usual enormous durations of time. Such a fluctuation might manifest itself as a flow of energy out of the black hole into a dispersed state, with a consequent reduction in the event horizon area, σ.

A closer inspection shows that a black hole fluctuation is not possible. Hawking's theorem (4.26) is on a different footing from the law of Clausius (2.3). The former is a genuinely *irreversible* law, even in the microscopic sense, whereas the latter is only correct to a large degree of probability. The origin of the genuine irreversibility in the black hole system is the event horizon, which ensures that once a particle has passed inside it, no amount of energy can enable it to leave again. Irrespective of the processes which may go on inside the horizon, all timelike world lines end on the singularity (figure 4.6).

Black hole irreversibility may be schematically represented by an analogue of Γ space, even though the non-relativistic treatment of statistical mechanics given in chapter 2 is not directly applicable here. Consider the region Ω in figure 4.7, in which the points represent microscopic states of the self-gravitating system. A representative point will move about in Ω as the system evolves, in the same fashion as for Γ space. Fluctuations in thermodynamic systems are assured by the recurrence theorem of Poincaré, which demands that a point will always return eventually to any region of the energy surface in Γ space, however low the entropy of that region. In Ω space the situation is quite different. There exists a region $\Omega' \supset \Omega$ with the property that a

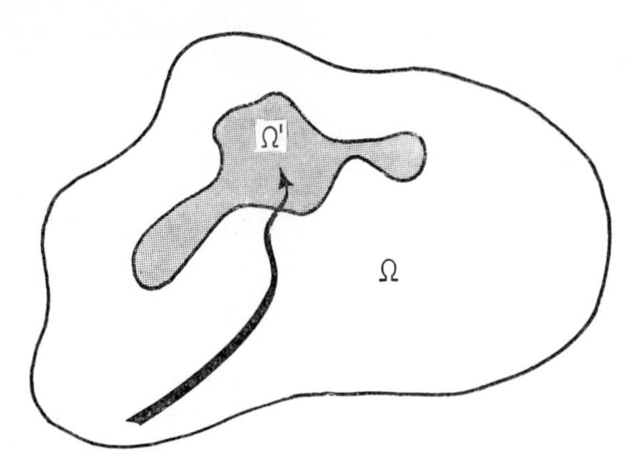

Figure 4.7 The region Ω is a highly schematic analogue of Γ space, applied to a relativistic self-gravitating system. Objects like the sun, stabilized outside their Schwarzschild radii, are represented by points which move in the unshaded region of Ω. If its pressure gradient is insufficient, and an object retreats inside its Schwarzschild radius, the trajectory of the representative point crosses into the shaded region Ω′, from which it can never emerge.

representative point in Ω′ cannot leave, even though it is allowed to enter. The boundary of Ω′ thus corresponds to the event horizon of the physical system. The existence of inaccessible regions of Ω space invalidates the Poincaré theorem; if a system begins at a point outside Ω′, gets into a collapse situation and crosses into Ω′, it cannot return to the former dispersed state outside the horizon.

At first sight the existence of this new type of irreversibility appears to conflict with the time reversal symmetry of the field equations (4.2), but this is erroneous. There is nothing to prevent the existence of a 'white hole'—the time reversed solution of the black hole collapse[21]. Indeed, the 'big bang' of the Friedmann cosmologies is an example of a white hole. If white holes were to occur as frequently as black holes there would be global time symmetry with irreversibility. On the other hand, it could well be imagined that the boundary conditions for each type of object are not equally favoured, so that the phenomenon of gravitational collapse supplies an additional asymmetry to our world. In any case, the reader should be careful to note that a black hole cannot turn into a white hole, but a white hole can turn into a black hole. In matter conserving cosmologies, this fact alone ensures some time asymmetry.

The above considerations may well be invalidated by a future quantum theory of gravity, which might permit quantum fluctuations—possibly even the 'radioactive decay' of black holes by a tunnel effect[22].

4.7 The big bang

As explained in section 4.1, the time asymmetric behaviour of local branch systems requires an overall thermodynamic disequilibrium in the universe. We have concluded that although the cosmological expansion is the cause of such a disequilibrium it is the thermonuclear production of starlight and the tendency for self-gravitating inhomogeneities to form that accounts for most of the conspicuous and significant entropy production in our region of the universe. Indeed, it is difficult to think of any terrestrial branch systems which do not ultimately depend for their formation on one of these two processes, whereas the expansion of the universe would not be expected to produce important branch systems here on Earth. It follows that an explanation of the asymmetry in time of the type of phenomena that we actually encounter as human beings must account for the existence of the latter two types of disequilibrium in the universe.

There have been three varieties of explanation proposed to explain the low entropy condition of the world from a cosmological standpoint. The first, due to Boltzmann[23], surmizes that we are living in a colossal Poincaré type fluctuation from the normal equilibrium condition of the world. The reason for our participation in such a staggeringly rare occurrence is attributed to the fact that the formation of biological matter itself requires the thermodynamic disequilibrium so produced; human beings could not exist to observe the equilibrium state. The conjecture can be faulted on several grounds. First, a fluctuation which produced the *present* low entropy condition of the universe is overwhelmingly more likely than one which produced a still *lower* entropy state in the past (this was fully explained in section 3.4). Yet there are non-thermodynamic reasons why we know that the entropy of the universe was lower in the past than its is now; for example, when distant galaxies are observed they are seen as they were many millions of years ago in a condition of thermodynamic disequilibrium. Another objection to Boltzmann's suggestion is that a fluctuation just on the size of the solar system would be sufficient to ensure the existence of life on Earth, and such a fluctuation is *far* more probable than one of cosmic proportions.

The second variety of explanation attempts to demonstrate that the thermodynamic disequilibrium in the world is re-established *continuously*, and is in some cosmological way *maintained*. One theory of this type is the steady state cosmology of Gold, Bondi and Hoyle, to be more fully discussed in section 7.2.

In the absence of continual creation of matter, it is necessary to consider the 'big bang' type of event, as in the Friedmann models discussed in section 4.2. Because of the finite age of these models, the entire universe can be garded as a sort of gigantic branch system, which was created in a low entropy state at $t = 0$ and is in the process of running through its course to equilibrium. According to this view, the continuing thermodynamic asymmetry observed in nature is attributed to the long relaxation times for stellar and gravitational processes, compared to the expansion time scale of the universe. As in the case of branch systems, the initial low entropy condition of the universe is not attributed to a fluctuation from an earlier equilibrium state, because the universe simply did not exist prior to this creation event. However, unlike the situation with branch systems, it is not possible to account for the low entropy initial state of the universe as due to interaction with the outside world, because the universe *is* the whole world. Therefore, at first sight one is apparently forced to accept the unsatisfactory conclusion that the negative entropy in the universe was simply 'put in' at the creation as an initial condition. Such an explanation is unlikely to impress anyone unless a reason can be found why the particular low entropy condition near the creation is an otherwise plausible one. Moreover, it is necessary to examine the thermodynamic properties of the early stages to determine the precise nature of the negative entropy.

At this point a word of caution should be added. On passing to earlier and earlier epochs, the density and energy of the cosmological fluid increases without limit according to the equations of general relativity. Our knowledge of the laws of physics can only be obtained up to finite energies by experimentation. However sophisticated our apparatus becomes we can never be sure that new physics will not apply in more extreme circumstances. At some stage in the past an analysis in terms of the known laws of physics breaks down and, if the singularity that occurs in the Friedmann models is a reality, everything (including relativity) breaks down there. At present, few people are confident of extrapolating the known laws of physics before about 10^{-23} seconds, when the density of the cosmological fluid is so large that the strong interaction between elementary particles is important. However, it will be seen that a knowledge of the very early stages is not relevant to an understanding of the origin of temporal asymmetry.

The subject of the asymmetry of starlight emission will be dealt with first. Starlight is produced in the hot stellar interiors by the nucleosynthesis of light nuclei to form stable heavy elements, with the attendant emission of energy in the form of photons and neutrinos. To trace the origin of this source of entropy producing photons, one must first ascertain the reason why the stars are made of light elements (mainly hydrogen and helium) in the first place. Iron is the most stable form of matter, so it might be asked why the universe is not already made entirely of iron. To answer this question requires a short digression into the subject of nucleosynthesis in the hot fireball of the 'big bang'.

Attention will be focussed on a small element of the cosmological fluid as we pass back in time towards the initial creation event. The temperature rises according to equations (4.16) and (4.18), until eventually the matter is found to be ionized when the temperature rises above about 4000 K. The matter and radiation are then strongly coupled and reach thermodynamic equilibrium; because the thermal capacity of the photons exceeds that of the baryons (protons and neutrons) by about 10^9, the matter will follow the radiation temperature as given by equation (4.16). At this epoch the dynamical state of the universe is also determined by the radiation, which is much more energetic than the matter. General relativity then gives the following relation between the temperature T and the time t in seconds since the creation event

$$T = \frac{10^{10} K}{t^{\frac{1}{2}}}. \tag{4.30}$$

At about $t = 1$ second the temperature is high enough for the photons to create electron–positron pairs. As the temperature continues to rise, more and more species of particles are produced this way, and all these species of particles (and the radiation) come into equilibrium with each other as the density rises. The equilibrium concentration for any particular species of elementary particle can be evaluated for a given temperature T. For example, at $T = 10^{10}$ K the ratio of neutrons to protons is about 15%. Now, for $t < 1$ second the statistical equilibrium between neutrons and protons depends on the weak interaction processes

$$\mathrm{n} + \mathrm{e}^+ \rightarrow \mathrm{p} + \bar{\nu}_e, \qquad \mathrm{p} + \mathrm{e}^- \rightarrow \mathrm{n} + \nu_e$$

which require the presence of electron–positron pairs. At $t \simeq 1$ second, these pairs disappear abruptly, and the n/p ratio becomes frozen at 15% for a few hundred seconds, after which it becomes appreciably reduced as a result of neutron decay. However, before this the neutrons will start to combine with the protons to form deuterium

$$\mathrm{n} + \mathrm{p} \rightarrow \mathrm{D} + \gamma \tag{4.31}$$

and provided the density is high enough this reaction will proceed so quickly that *all* the neutrons will be incorporated in deuterium. This key reaction will not proceed, however, until the temperature has fallen below 10^9 K (at $t \simeq 100$ seconds) because the energetic photons immediately disintegrate the newly formed deuterons; that is, before 10^9 K, free neutrons and protons represent the maximum entropy state, but afterwards deuterons are the equilibrium form of matter.

As soon as the reaction (4.31) is complete, further rapid reactions occur to form helium. The abundance of helium in the universe is therefore determined by the frozen-in ratio at $T = 10^{10}$ K. With a 15% n/p ratio, a helium/hydrogen ratio around 30% is expected. Detailed calculations indicate about 27%. The fact that the measured value of this ratio is betwee 25% and 30% is remarkable confirmation of the big bang theory.

As regards the heavier elements, these calculations show that only minut proportions are built up during the initial fireball, because the reaction speeds which depend on the particle densities and the temperature, are falling rapidly and there is insufficient time for an appreciable amount of these elements t form.

Of course, it is crucial for our explanation of the large amount of hydroge in the universe that the uncombined protons do not all burn up by the p– reaction, or they would not survive to undergo nucleosynthesis in stars. Thi does not occur due to the extremely long process time for this reaction. It i interesting that this result depends delicately on the value of the stron coupling constant. If it were a mere few per cent larger it would be possibl for two protons to form a bound di-proton, leading very rapidly to deuteriun and then helium.

The origin of the other major source of temporal asymmetry in the worl will now be considered, namely, the growth of gravitational density perturba tions. A moment's reflection will show that, for example, the terrestrial tide are a result of this phenomenon. As discussed in section 4.6, self-gravitatin systems have no true equilibrium configuration. Whereas a flask of gas in th laboratory will soon reach a condition of uniform temperature, pressure an chemical composition, from which it will not change, if a large enough volum of cosmological fluid is in such a uniform state it will begin to undergo larg density fluctuations, possibly even terminating in gravitational collapse.

It is most important for the reader to distinguish between the *local* depar tures from homogeneity due to gravitational perturbations and the *globa* motion of the homogeneous cosmological fluid, which is also determined b the gravitational field equations (4.2). In the former case the collapsing flui will experience large entropy increases, with the establishment of temperatur and pressure gradients, turbulence, etc, but in the latter case there is, as al ready seen, no change in the entropy of a comoving volume (to a goo approximation) because of the homogeneity.

From the cosmological point of view it is most satisfactory that one ma expect the accumulation of structure in the universe on thermodynami grounds[25]. The present condition of the universe is a hierarchy of structure— the planets, the solar system, star clusters, galaxies, clusters of galaxies. It i consequently an attractive possibility that the world began without thi structure, as a simple uniform fluid, and that these local departures fron homogeneity grew from the natural fluctuations in the fluid occurring durin the early epochs after the big bang.

The features of the growth of density perturbations in the Friedmanr models, first described by Lifshitz[26], will now be summarized. Neglecting pressure for the moment, recall the three solutions (4.9) representing the matter filled Friedmann models. For small R, $\eta \ll 1$ and all three solutions for R only differ by a term of order η^4. If a spherical region of cosmological fluid is slightly perturbed so that it contains matter expanding more slowly than elsewhere, this perturbed region will correspond to a cosmological model with lower energy. Because such lower energy models expand more slowly, this region will lag more and more behind the rest of the universe, and the perturbation will grow. If the scale factor for this region is changed by ΔR, then $\Delta R \propto \eta^4 \propto R^2$, so

$$\frac{\Delta R}{R} \propto \frac{\Delta \varepsilon}{\varepsilon} \propto R \propto t^{\frac{2}{3}} \tag{4.32}$$

i.e. the density perturbations grow proportional to a power of t. This result is valid even when the size of the region is larger than the particle horizon.

In practice, pressure cannot be neglected, especially at early epochs when the universe is radiation dominated. The effect of pressure gradients is to stabilize the density perturbations on a scale $\lessdot c_s/(G\rho)^{\frac{1}{2}}$, where c_s is the sound speed. This is known as the Jeans' length, and any perturbation less than this will not grow but oscillate[27]. During the radiation dominated era $c_s = c/\sqrt{3} = 1/\sqrt{3}$ (c = speed of light = 1), so that the Jeans' length is comparable with the particle horizon distance. Perturbations on a larger scale than the horizon will continue to grow until they come within the horizon, after which they will start to oscillate.

It was originally hoped that random statistical fluctuations of order '$\sqrt{N}$' would be sufficient to account for the eventual existence of galaxies and galactic clusters. Unfortunately, because of the slow (power law) growth rate of these fluctuations, due to the expansion, they are far too small to account for galactic sized objects unless initiated at a very early stage, when the particle horizon contained only a few atoms! At the present state of the theory it appears that appropriate amplitude perturbations must simply be postulated as initial conditions to explain the existence of galaxies.

From the point of view of thermodynamic irreversibility, we are more interested in the small scale perturbations which eventually form stars and planets. These perturbations are stabilized by radiation pressure until decoupling of matter and radiation (about 10^3 years), after which the matter becomes transparent to photons, and the Jeans' length drops abruptly (about a factor of 10^7) because the matter is only stabilized by the thermal pressure. Thereafter, the Jeans' length falls steadily, permitting the formation of stars, etc.

What conclusion can be drawn from this short digression into the condition of the cosmological fluid at early epochs? Provided the discussion is restricted to sufficiently small volumes, it is fairly clear that the cosmological fluid was in a state of *local* thermodynamic equilibrium at early epochs. The word 'equilibrium' used in this context needs careful qualification in two ways (quite apart from the qualification mentioned at the end of section 3.1).

1 Focussing attention once again on a small element of fluid as we pass back to earlier and earlier times, a series of thresholds are reached at which more and more species of particles come into equilibrium with each other. For example, before $t = 10^3$ years, matter and radiation are coupled and in equilibrium; before $t = 10^{-4}$ seconds, neutrinos come into equilibrium with the matter, etc. As the energy rises, new types of particle are created and reach equilibrium with the others. Indeed, there is recent speculation[28] that this process may continue without limit as one passes nearer and nearer to the initial singularity, i.e. there are an infinite number of thresholds and an infinite number of new particles to be created. At any one time the word 'equilibrium' can only refer to a limited number of fluid components which happen to be coupled together at that energy. When the temperature drops below the relevant threshold, most of these particles rapidly decay to form stable baryons, leptons or photons, so they are of no consequence to the subject of negative entropy reservoirs at our own epoch. However, the later thresholds *are* important in this respect, because they involve stable particles that have persisted until today. When these stable fluid components decouple, an energy imbalance gradually develops between them as the universe expands, e.g. between matter and radiation, which cool at different rates. This would not be significant in the case of, say, neutrinos with regard to laboratory systems, the thermodynamic properties of which may be successfully described without allusion to their continual penetration by these essentially non-interacting particles. However, there are local special regions (stars) where matter becomes strongly coupled again both to electromagnetic radiation and neutrinos, the former of which is so important for everyday thermodynamic irreversibility.

At first sight it appears paradoxical that an element of the cosmological fluid can start out in a quasi-equilibrium condition, and yet still increase its entropy at later epochs, by starlight emission, etc. This paradox is resolved by an inquiry into the second qualification of the word 'equilibrium'.

2 From the discussion of section 4.6, it will be recalled that a self-gravitating system has no true equilibrium configuration. It therefore possesses an infinite reservoir of negative entropy. In the case of the cosmological fluid this is significant in two separate ways. The first is the possibility of local departures from homogeneity due to gravitational growth of density perturbations. In the early stages a small region of the fluid will be stabilized against collapse by pressure gradients. As the temperature drops, these pressure gradients fail, and the system collapses, perhaps to be stabilized in a new metastable condition at greater density. During this collapse part of the unlimited reservoir of the system's negative entropy is used up as the system undergoes an entropy increase. The second point is that the cosmological fluid is partaking of the general cosmological expansion, and so finds itself in a changing gravitational field on the global scale. The increase of entropy through irreversible processes such as starlight may be thought of as being 'paid for' by the gravitational field of the universe in a fashion that will be explained in greater detail in section 7.3. The self-gravitating universe as a whole has unlimited capacity to increase its entropy.

We have reached a remarkable conclusion. The origin of *all* thermodynamic irreversibility in the real universe depends ultimately on gravitation. *Any* gravitating universe that can exist and contains more than one type of interacting material *must* be asymmetric in time, both globally in its motion, and locally in its thermodynamics (see also chapter 7).

References

1. H. Bondi, *Cosmology*, Cambridge University Press, Cambridge, 1960.
2. A. Friedmann, *Z. Phys.*, **10,** 377, 1922; **21,** 326, 1924.
3. A. R. Sandage, *The Observatory*, **88,** 91, 1968; *Astrophys. J.*, **152,** 149, 1968.
4. A. A. Penzias and R. W. Wilson, *Astrophys. J.*, **142,** 419, 1965.
5. E. Hubble, *Proc. Nat. Acad. Sci.*, **15,** 168, 1929.
6. R. C. Tolman, *Relativity Thermodynamics and Cosmology*, Clarendon Press, Oxford, 1934, sections 170, 171.
7. R. C. Tolman, *Relativity Thermodynamics and Cosmology*, Clarendon Press, Oxford, 1934, section 122.
8. R. C. Tolman, *Relativity Thermodynamics and Cosmology*, Clarendon Press, Oxford, 1934, section 130.
9. Nevertheless, this seems to have been taken as the origin of *all* irreversibility by D. Layzer, *IUPAC International Conference on Thermodynamics*, Cardiff (Ed. P. T. Landsberg), Butterworth, London, 1970.
10. H. W. M. Olbers, *Bode Jb.*, **110,** 1826; *Edinburgh New Phil. J.*, April–October, 141, 1826.
11. T. Gold, *Amer. J. Phys.*, **30,** 403, 1962. H. Bondi, *The Observatory*, **82,** 133, 1962.
12. M. J. Rees, *The Observatory*, **89,** 193, 1969.
13. For example: L. D. Landau, *Phys. Z. Sowjet.*, **1,** 285, 1932. I. Prigógine and G. Severne, *Physica*, **32,** 1376, 1966.
14. W. C. Saslaw, *Mon. Not. Roy. Astr. Soc.*, **141,** 1, 1968; **143,** 437, 1969.
15. L. D. Landau, *Phys. Z. Sowjet.*, **1,** 285, 1932. S. Chandrasekhar, *Mon. Not. Roy. Astr. Soc.*, **95,** 207, 1935. J. R. Oppenheimer and G. Volkoff, *Phys. Rev.*, **55,** 374, 1939.

16. This a consequence of the so-called Israel–Carter conjecture. References to work which supports this conjecture can be found in J. D. Bekenstein, PhD Thesis, Princeton University, 1972. See also R. Ruffini and J. A. Wheeler, *Physics Today*, **24,** 30 January 1971.
17. C. W. Misner, J. A. Wheeler, private communications; J. D. Bekenstein, PhD Thesis, Princeton University, 1972, pp. 108, 131.
18. R. Penrose, *Rivista del Nuovo Cimento*, ser. 1, **1,** 252, 1969. D. Christodoulou and R. Ruffini, *Phys. Rev.*, D, **4,** 3552, 1971.
19. S. W. Hawking, *Comm. Math. Phys.*, **25,** 152, 1972.
20. J. D. Bekenstein, PhD Thesis, Princeton University, 1972.
21. The phrase 'white hole' seems first to have been used by R. Penrose at the Fifth Texas Symposium on Relativistic Astrophysics, December 1970. See also the remarks in B. K. Harrison, K. S. Thorne, M. Wakano and J. A. Wheeler, *Gravitation Theory and Gravitational Collapse*, University of Chicago Press, Chicago & London, 1965, chapter 11.
22. B. K. Harrison, K. S. Thorne, M. Wakano and J. A. Wheeler, *Gravitation Theory and Gravitational Collapse*, University of Chicago Press, Chicago and London, 1965, chapter 11.
23. L. Boltzmann, *Nature*, **51,** 413, 1895.
24. R. V. Wagoner, W. A. Fowler and F. Hoyle, *Astrophys. J.*, **148,** 21, 1967.
25. In spite of the fact that gravitating systems tend to grow structure thermodynamically, additional special principles have been invoked by D. Layzer, A unified approach to cosmology, in *Lectures in Applied Mathematics* (Ed. J. Ehlers), vol. 8, American Mathematical Society, Providence, R. I., p. 237. See also The strong cosmological principle, indeterminacy and the direction of time, in *The Nature of Time* (Ed. T. Gold), Cornell University Press, Ithaca, 1967 and the remarks by P. T. Landsberg, *Studium Generale*, **23,** 1128, 1970.
26. E. Lifshitz, *Zurn. Eksp. Teor. Fis.*, **10,** 116, 1946. Review articles have been given by E. R. Harrison, *Rev. Mod. Phys.*, **39,** 862, 1967 and M. J. Rees, Some current ideas on galaxy formation, in *General Relativity and Cosmology* (Ed. R. K. Sachs), Academic Press, New York, 1971.
27. J. H. Jeans, *Astronomy and Cosmogony*, Cambridge University Press, Cambridge, 1928.
28. R. Hagedorn, *Suppl. Nuovo Cimento*, **3,** 147, 1965; *Nuovo Cimento*, **56**A, 1027, 1968.

Further reading

1. Clear accounts of the thermodynamic aspects of cosmology may be found in R. C. Tolman, *Relativity Thermodynamics and Cosmology*, Clarendon

Press, Oxford, 1934, and D. W. Sciama, *Modern Cosmology*, Cambridge University Press, Cambridge, 1971.

2. Some authors have expressed doubts about the application of the entropy concept to the whole universe. In this connection see K. P. Stanyukovic, On the increase of entropy in an infinite universe, *Dokl. Akad. Nauk. SSSR*, **69**, 793, 1949; L. Tisza, *Math. Rev.*, **12,** 787, 1951.
3. A recent review article on black holes is given by R. Ruffini and J. A. Wheeler, Introducing the black hole, *Physics Today*, **24,** 30, January 1971. For an explanation of why collapsing stars may be treated as 'true' black holes even after a finite time see Why 'black hole'?, in C. W. Misner, K. S. Thorne and J. A. Wheeler, *Gravitation*, Freeman, San Francisco, 1973, section 33.1. See also section 33.8 Reversible and irreversible transformations of black holes.
4. For recent correspondence concerning entropy and black holes see J. D. Bekenstein, *Lettere al Nuovo Cimento*, **4,** 737, 1972; L. Basano and A. Morro, *Lettere al Nuovo Cimento*, **6,** 193, 1973; W. Israel, *Lettere al Nuovo Cimento*, **6,** 267, 1973. See also C. W. Misner, K. S. Thorne and J. A. Wheeler, *Gravitation*, Freeman, San Francisco, 1973.
5. A dissenting attitude to some of the problems raised in this chapter are discussed by K. Popper, Time's arrow and entropy, *Nature*, **207,** 233, 1965. See also W. Buchel, Entropy and information in the Universe, *Nature*, **213,** 319, 1967; B. Gal-Or, Are the astrophysical and statistical schools of irreversibility compatible?, *Nature*, **234,** 217, 1971; J. Merleau-Ponty, Temps et cosmogonie, in *Cosmologie du XXe Siècle*, Gallimard, Paris, 1965.

5 Electromagnetic Waves

5.1 Retarded and advanced fields

The last three chapters have been devoted exclusively to understanding the nature and origin of temporally asymmetric and irreversible phenomena involving entropy and thermodynamics. The reader may have wondered whether there exist *non-thermodynamic* phenomena which behave asymmetrically in time. There does indeed exist an important and familiar class of processes which, although they may be accompanied by thermodynamic irreversibility, seem to display an additional element of asymmetry on their own.

As a first example, if I throw a stone into the centre of a pond, the result that I expect to see is a diverging series of circular waves, spreading outwards from the region of impact towards the edge of the pond. I do *not* expect to see waves leaving the edge and converging onto the point of the stone's impact at the moment of contact (see figure 5.1). Likewise, if I send a message through a radio transmitter to a distant colleague, or if I shout to him across the intervening space, I do not expect him to know my message *before* it is sent, but naturally presume that the radio or sound waves will travel outwards from the transmitter or my mouth, to reach a distant point at a later time. And by 'earlier/later' I mean in the direction of entropy increase.

Mathematically, all three of the above physical processes share the property of being *waves*: water surface waves, radio waves, sound waves. The well-known equation for wave propagation in flat space is

$$\left(\frac{1}{c^2}\frac{\partial^2}{\partial t^2} - \nabla^2\right)\phi(\boldsymbol{r}, t) = 4\pi\rho(\boldsymbol{r}, t). \tag{5.1}$$

ϕ is the amplitude of the wave disturbance at the space–time point $(\boldsymbol{r}, t)$, ρ is the corresponding source density and c the wave velocity. For simplicity, at first only non-dispersive, non-absorptive media will be considered, so that c is a constant.

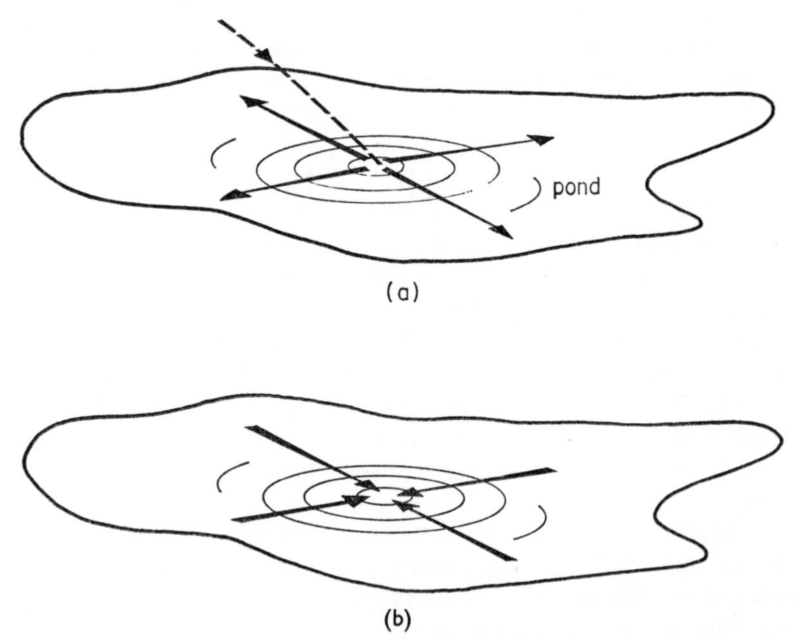

Figure 5.1(a) A pattern of circular waves diverging from a central point of disturbance, such as might be caused by the impact of a stone. (b) The time reversed situation, with coherent circular waves converging spontaneously towards a central point. This phenomenon never seems to be observed.

Equation (5.1) has a well-known solution[1] in three dimensions

$$\phi_{ret}(\boldsymbol{r}, t) = \int_{\substack{\text{all space–}\\ \text{time}}} \frac{\delta\left(t' + \dfrac{|\boldsymbol{r} - \boldsymbol{r}'|}{c} - t\right)}{|\boldsymbol{r} - \boldsymbol{r}'|} \rho(\boldsymbol{r}', t')\, d^3r'\, dt'. \tag{5.2}$$

δ is the Dirac δ function which has the following property. It enables us to carry out the t' integral in (5.2) to obtain

$$\phi_{ret}(\boldsymbol{r}, t) = \int_{\text{all space}} \left[\frac{\rho(\boldsymbol{r}', t')}{|\boldsymbol{r} - \boldsymbol{r}'|}\right]_{ret} d^3r' \tag{5.3}$$

where the bracket $[\ ]_{ret}$ means that the time t' is to be evaluated as the *retarded* time $t - |\boldsymbol{r} - \boldsymbol{r}'|/c$. This retardation expresses the fact that the disturbance at position $\boldsymbol{r}$ at time t is caused by the source ρ at another point $\boldsymbol{r}'$, but not at a simultaneous time t; instead at an earlier time t', the difference being accounted for by the delay caused while the disturbance propagates across the intervening distance $|\boldsymbol{r} - \boldsymbol{r}'|$ at the velocity c. The total disturbance $\phi_{ret}(\boldsymbol{r}, t)$ is the linear superposition of all these earlier sources, given by (5.3).

However, another solution to equation (5.1) may be obtained by reversing the sign of $|\boldsymbol{r} - \boldsymbol{r}'|/c$ in (5.2). This solution may be written

$$\phi_{adv}(\boldsymbol{r}, t) = \int \left[\frac{\rho(\boldsymbol{r}', t')}{|\boldsymbol{r} - \boldsymbol{r}'|}\right]_{adv} d^3r' \tag{5.4}$$

where the bracket $[\ ]_{adv}$ means that t' must be evaluated at the *advanced* time $t + |\boldsymbol{r} - \boldsymbol{r}'|/c$. Physically, this corresponds to a situation where the disturbance at $(\boldsymbol{r}, t)$ is due to the sources at points $\boldsymbol{r}'$, and at times t' that lie in the *future* of t. It should be noted that (5.4) is not the time reverse of (5.3), but the time reverse with a time reversed source ρ. The first (retarded) solution correctly describes the situation illustrated by the examples cited above, whilst the second (advanced) solution does not seem to be observed in nature. It would correspond, for instance, to a radio wave coming from infinity and converging onto a radio transmitter.

The difference between (5.3) and (5.4) is attributed to the *boundary conditions*. Equation (5.1) is a second order hyperbolic partial differential equation. From any solution another may always be obtained by adding a solution of the homogeneous (source free) equation

$$\frac{1}{c^2}\frac{\partial^2\phi}{\partial t^2} - \nabla^2\phi = 0. \tag{5.5}$$

It may be shown[2] from the theory of hyperbolic equations that in order to obtain a unique, stable solution to (5.1) the appropriate boundary conditions are a specification of ϕ and $\dfrac{\partial\phi}{\partial t}$ throughout all space at one time t. In general, we may match to any boundary conditions of this sort by adding a suitable solution of (5.5) to (5.3) or (5.4). For example, the boundary conditions appropriate to solution (5.3) are $\phi = \dfrac{\partial\phi}{\partial t} = 0$ for $t < 0$, whilst for (5.4) $\phi = \dfrac{\partial\phi}{\partial t} = 0$, $t > 0$. The difference between these solutions, $\phi_{ret} - \phi_{adv}$, is therefore a solution of the homogeneous equation (5.5).

These ideas are better illustrated by replacing the differential equation (5.1) by an integral representation. It may be shown[3] that

$$\phi(\boldsymbol{r}, t) = \int_V \frac{[\rho]_{ret}}{R}\, dV + \frac{1}{4\pi}\int_S \left\{[\phi]_{ret}\frac{\partial}{\partial n}\left(\frac{1}{R}\right) - \frac{1}{R}\frac{\partial R}{\partial n}\left[\frac{\partial\phi}{\partial t}\right]_{ret} - \frac{1}{R}\left[\frac{\partial\phi}{\partial n}\right]_{ret}\right\} dS \tag{5.6}$$

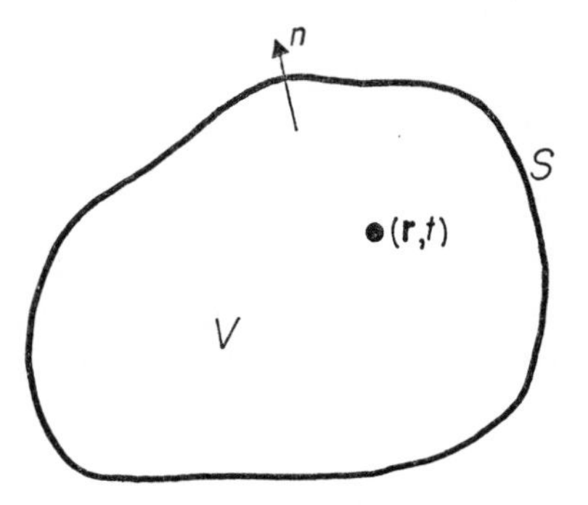

Figure 5.2 A surface S enclosing a volume V including a general space–time point. n is an outward normal to S

where $c = 1$ for convenience (in our units this is already so for electromagnetic waves) and $R = |\boldsymbol{r} - \boldsymbol{r}'|$. The meaning of the integrals in (5.6) is illustrated in figure 5.2, which depicts a volume V of Euclidean space bounded by a smooth closed surface S, having an outward normal denoted by n. The position vector $\boldsymbol{r}$ refers to a point inside V. The contribution to the total field ϕ at $(\boldsymbol{r}, t)$ is attributed to three origins:

(1) sources inside V; these are accounted for by the first (volume) integral,
(2) sources outside V,
(3) source free disturbances coming from infinity.

The latter two contributions are contained in the second (surface) integral. The first integral satisfies the inhomogeneous wave equation and the second integral satisfies the homogeneous equation, inside V.

Another possible integral representation of the *same field* may be obtained by replacing (5.6) by a similar expression, with $[\ ]_{ret}$ replaced by $[\ ]_{adv}$ throughout. Such an expression will be abbreviated as follows:

$$\phi = \int_V adv + \int_S adv \tag{5.7}$$

while (5.6) could be abbreviated

$$\phi = \int_V ret + \int_S ret. \tag{5.8}$$

Because of the linearity of the wave equation, any linear combination of *ret* and *adv* may be taken

$$k\int_V ret + (k-1)\int_V adv + k\int_S ret + (k-1)\int_S adv \tag{5.9}$$

($k < 1$); in particular,

$$\tfrac{1}{2}\int_V (ret + adv) + \tfrac{1}{2}\int_S (ret + adv). \tag{5.10}$$

If it is now supposed that the sources of ϕ are located in a small region within V, then the contribution to the surface integrals arises only from source free disturbances from infinity. In order to recover the usual retarded solution (5.3) from (5.8) it is necessary to put

$$\int_S ret = 0 \tag{5.11}$$

that is, require that the source free radiation in the retarded case vanish also. On the other hand, to recover equation (5.3) from the advanced form (5.7), the source free radiation cannot be allowed to vanish, as this would lead to $\phi = \int_V adv$. Indeed

$$\int_S adv = \int_V ret - \int_V adv \tag{5.12}$$

which does not in general vanish[4].

Thus, in a discussion of wave propagation, we are equally allowed to use retarded or advanced formulations. However, the boundary conditions in either case have to be chosen differently, so that there is no source free radiation coming into the region of interest from the remote past, although there are disturbances propagating outwards from the region of interest into the remote future. We shall now try to understand why this choice of boundary conditions is always made in nature.

5.2 Waves in finite systems

There appears to be a close analogy between a closed finite isolated medium, undergoing wave disturbances of a given energy, and the Boltzmann model of an isolated box of gas treated in chapters 2 and 3. As an example of this, consider an idealized model of a 'pond' of water, delineated by an irregular shaped boundary, which has the property of perfectly reflecting the surface waves of the 'pond'. All damping effects such as viscosity and friction at the boundary will be neglected, because these represent thermodynamic effects which are only incidental to the asymmetry under discussion. It will be assumed also that the system is isolated, so that the surface will not be disturbed except at an initial instant when a stone is thrown into the centre.

If the surface was undisturbed before the stone was thrown, the resulting wave pattern will be a diverging series of circular waves, spreading outwards from the region of impact towards the irregular boundary, whereupon reflections will occur in a highly complicated and apparently haphazard way. After a few such reflections the surface of the 'pond' will take on a more or less uniform appearance, with small wavelets growing or dying in a complex way as differing wave patterns interfere. The wave interference is analogous to the collisions of gas molecules in Boltzmann's model system, and the progress from a simple symmetric wave pattern (which may be described by a few parameters) to a complex distribution of wavelets (which requires a great deal of information to describe) closely resembles the progress of a gas from a low entropy to a high entropy state; in particular, the explosive expansion of a confined gas into a larger vacuum.

It is tempting to exploit this analogy still further, and try to account for the asymmetry of retarded wave motion along the lines of chapters 2 and 3 for entropy change. Thus, it might be conjectured that the wave system was in some sense *ergodic*; that any arbitrary surface configuration compatible with the total energy of the disturbance would eventually be realized to within arbitrary accuracy. The proof of such a conjecture might depend critically on how the 'closeness' of two configurations was defined. Of course, it would be necessary to avoid the analogue of 'exceptional trajectories' which occur in ergodic theory under extremely symmetric conditions (see p. 48) by avoiding symmetrically shaped 'ponds' (for example, a circular pond disturbed at the centre could only produce circular wavefronts).

If such a 'wave ergodic' hypothesis were true, we should then have a natural interpretation of retarded propagation in terms of *branch systems*. In a completely isolated 'pond' the water surface would be in a uniform state of confusion most of the time, corresponding to equilibrium. Very rarely though, a large fluctuation would occur. Sometimes this would take the form of a converging circular disturbance, which would be interpreted as 'advanced wave motion', and corresponds to the equally unfamilar entropy decreasing fluctuations in a gas. On reaching the pond's centre, this 'advanced wave' would pass on through and become a retarded outgoing circular wave, similar to that which the stone's impact caused. This would be a Poincaré type recurrence. Thus, for every set of motions interpreted as retarded waves there is another set, occurring equally often, interpreted as advanced waves.

However, a real pond would not be completely isolated for all time, but would instead be a branch system in the following sense. The surface of the pond would not contain a disturbance of interest *before* the stone was thrown. (Should the pond contain some sort of prior disturbance, this would not matter provided the branch system was formed at *random*, i.e. the stone thrown at random, for then it would be overwhelmingly improbable that the

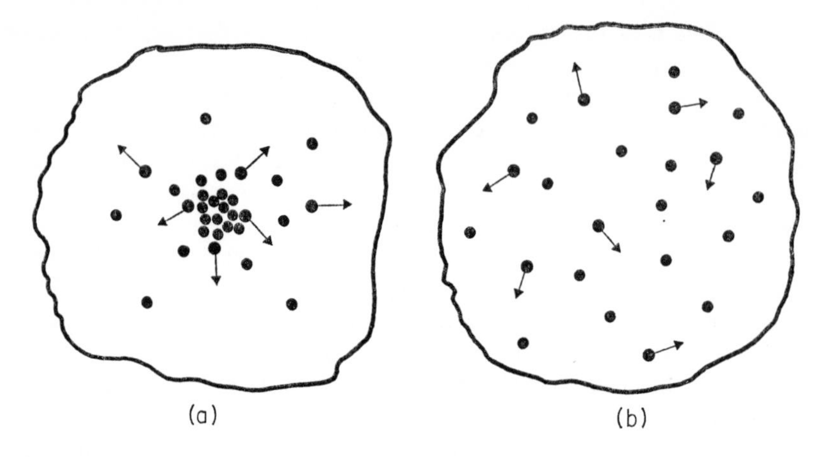

Figure 5.3 Expansion of a gas into a vacuum. In (a) the gas molecules are released at the centre of a large container, and stream outwards in a diverging symmetric disturbance. After the disturbance has reached the walls of the container, the molecules soon settle down into a high entropy, disordered configuration such as that shown in (b), though Poincaré's theorem assures that they will return to the centre eventually. The similarity of (a) to the surface waves shown in figure 5.1 is obvious.

existing wavelets would combine at the appropriate moment in such a manner that the total superposed disturbance became a contracting wave). In this way the existence of retarded waves is assured by the asymmetric way in which the branch system is formed. The situation is strikingly similar to the behaviour of a compressed gas when allowed to expand into a vacuum in a larger container (see figure 5.3). The molecules constitute a spreading spherical disturbance until they reach the walls of the container, after which interparticle collisions between reflected molecules and others soon reduce the gas to a uniform, chaotic, high entropy condition. The similarity is perhaps quite understandable when it is remembered that the gas molecules should really be described by de Broglie waves in quantum mechanics.

In section 3.4 it was explained how a real thermodynamic system is not isolated for two distinct reasons: (1) branch systems form in the first place by interaction with the outside, (2) there is always a residual continuous coupling with the outside world through the walls of the box. It was also carefully explained how the origin of the entropic asymmetry there displayed is already present through (1), and does not require (2), although the latter process does have the effect of destroying the microscopic reversibility of the system. In the present case there is also a residual, continuous coupling of the system to the outside world which is only incidental to the existence of retarded waves, these already being accounted for by the asymmetry of the branch system formation. Nevertheless, this residual coupling does destroy the reversibility of the system by creating random 'noise', which obliterates the initial conditions. In practice, this coupling can be in the form of frictional and viscous effects, or incidental sources of disturbance such as the wind. These effects could be included by adding new source terms to the wave equation (5.1) which, by the nature of their origin (thermodynamic processes) are not symmetric under time reversal. Physically this 'noise' means that a simultaneous reversal of all the wavefront's directions of propagation would no longer cause the system to evolve through its past history in reverse sequence (which it would in the case of complete isolation). Nevertheless, it is concluded that the existence of retarded waves is assured independently of considerations of thermodynamics.

This section is concluded with some remarks about *correlations* and *coherence*. Returning yet again to the case of a gas confined in a box, it will be recalled that the stochastic motion of the walls, due to random outside influences, destroys the correlations between the molecules, thus bringing about an increase in the fine-grained entropy. Molecules which approach a particular region from different directions will have undergone random collisions with different parts of the wall surface, and will be uncorrelated.

The hypothesis that influences emanating from different directions in space are uncorrelated has been framed as a general law, called the *law of conditional independence*, by Penrose and Percival[5]. Its plausibility lies in the fact that the existence of a particle horizon ensures that distant regions of space lying in different directions are causally unconnected. The relevance of this law to wave propagation is the following. Waves on a real pond are usually damped away at the edges by frictional effects. The reverse process, in which the spontaneous motion of the particles at the edges combine favourably to bring about the generation of a disturbance is overwhelmingly improbable, though not impossible, on thermodynamic grounds. Nevertheless, *in spite of this fact*, the spontaneous production of a wave by the pond's edges would very probably not produce a converging, *coherent* circular wavefront, precisely because this would require the cooperation of influences emanating from different directions of the edges. This is an expression of the fact that there is an *additional* reason why advanced waves should not occur, independent of thermodynamic considerations, although thermodynamic irreversibility does accompany the process of wave progagation. This point has been stressed by Popper[6], who draws attention to the fact that a coherent converging circular wave on a water surface would require *organization from the centre* (e.g. by dropping a circular ring horizontally onto the water surface), but would be well nigh impossible spontaneously[7].

Although the remarks of this section have been applied to the example of surface water waves, they apply quite generally to all types of waves in *finite* systems. When generalizing the analysis to include *infinite* (i.e. unbounded) systems, problems occur, because there may be source free disturbances from infinity, and because the edges of the system cannot reflect back the wavelets to form a chaotic condition. The same problems crop up in thermodynamics, when the gas under consideration is no longer confined to a finite volume. It is not then possible to apply ergodic theory, the recurrence theorem, etc, because these depend upon the finitude of the system for their validity. A hot gas released into infinite space (neglecting gravity) will expand away for ever. This is a time asymmetric process, but of a clearly different character from that found by entropic considerations. One way of eliminating the possibility of converging infinite waves, or clouds of gas, is to simply accept the law of conditional independence ad hoc. However, various authors have attempted to account for this asymmetry on the basis of the known laws of physics.

5.3 Basic electrodynamics in Minkowski space

The most relevant system that permits the propagation of disturbances to infinity is the electromagnetic field. Conventional Maxwell–Lorentz electrodynamics treats two mechanical systems. The electromagnetic field is the first, the measurable quantities being the electric field strength $\boldsymbol{E}(x)$ and the magnetic induction $\boldsymbol{B}(x)$. For relativistic purposes it is usual to combine the six components of these two vectors into an antisymmetric second rank field strength tensor $F^{\mu\nu}$ defined by

$$F^{\mu\nu}(x) = \begin{bmatrix} 0 & -E_1 & -E_2 & -E_3 \\ E_1 & 0 & -B_3 & B_2 \\ E_2 & B_3 & 0 & -B_1 \\ E_3 & -B_2 & B_1 & 0 \end{bmatrix}. \tag{5.13}$$

Often it is more convenient to introduce the unobservable four-potential $A^{\mu}(x)$, in terms of which $F^{\mu\nu}$ is given by

$$F^{\mu\nu} = A^{\mu,\nu} - A^{\nu,\mu} \tag{5.14}$$

$\left(,\mu \text{ is shorthand notation for } \dfrac{\partial}{\partial x_{\mu}}\right)$. The additional freedom acquired by the introduction of a potential is sometimes removed by imposing the so-called Lorentz condition

$$A^{\mu}{}_{,\mu} = 0. \tag{5.15}$$

The second mechancial system of concern is a collection of charged particles which act as sources for the fields. Let the ith particle have mass m_i and charge e_i. Its trajectory in space–time is given parametrically by the equation

$$x^{\mu}_{(i)} = z^{\mu}_{(i)}(\tau_i) \tag{5.16}$$

τ_i being the ith particle proper time. $\dot{z}^{\mu}_{(i)}$ is the four-velocity of this particle, which may be used to construct a conserved current vector for point charges

$$j^{\mu}(x) = \sum_i e_i \int_{-\infty}^{\infty} \delta^4(x - z_i) \dot{z}^{\mu}_{(i)} \, d\lambda_i \tag{5.17}$$

from which it follows that

$$j^{\mu}_{,\mu} = 0 \tag{5.18}$$

which is the equation of current conservation. The dot denotes total differentiation with respect to the parametric argument of z. In the integral (5.17) this is an arbitrary parameter λ_i which is monotonically increasing along the particle's world line; λ_i is normally identified with τ_i at the end of a calculation. δ^4 is the four dimensional Dirac δ function, i.e. $\delta(x_1)\,\delta(x_2)\,\delta(x_3)\,\delta(x_4)$. It is the result[8] of operating on $\frac{1}{4\pi}\,\delta(x^{\mu}x_{\mu})$ with the d'Alembertian

$$\Box^2\, \delta(x^{\mu}x_{\mu}) = 4\pi\, \delta^4(x). \tag{5.19}$$

(This operator may be written $\eta^{\mu\nu} \dfrac{\partial}{\partial x^{\mu}} \dfrac{\partial}{\partial x^{\nu}}$, and is the four dimensional Minkowski space analogue of the Laplacian ∇^2. It is a Lorentz covariant scalar.)

The equations of motion of the coupled matter–field system may be formally derived from a relativistic variational principle. The total action $\mathcal{S}$ may be written as a sum of three terms

$$\mathcal{S} = -\sum_i m_i \int (\dot{z}^{\mu}_{(i)} \dot{z}_{(i)\mu})^{\frac{1}{2}}\, d\lambda_i - \frac{1}{16\pi} \int F^{\mu\nu}F_{\mu\nu}\, d^4x - \sum_i e_i \int j^{\mu}A_{\mu}\, d\lambda_i. \tag{5.20}$$

The first term represents the inertial properties of all the particles, the second term the free motion of the electromagnetic field, and the third the coupling between the two systems. By varying the potentials A_{μ} and keeping the world lines z fixed, the principle of stationary action, $\delta\mathcal{S} = 0$, yields the equation of motion of the field—Maxwell's equations

$$\Box^2 A^{\mu} = 4\pi j^{\mu}. \tag{5.21}$$

Similarly, the zs may be varied with fixed A_{μ} to yield the particle equations of motion

$$m_i \ddot{z}^{\mu}_{(i)} = e_i F^{\mu}{}_{\nu}(z_i) \dot{z}^{\nu}_{(i)}. \tag{5.22}$$

If equation (5.22) is written in three-vector form it becomes

$$m_i \ddot{\boldsymbol{z}}_{(i)} = e_i(\boldsymbol{E} + \boldsymbol{v}_{(i)} \times \boldsymbol{B}) \tag{5.23}$$

which is the familiar Lorentz force law, with $\boldsymbol{v}_{(i)} = \dot{\boldsymbol{z}}_{(i)}$.

From the structure of the d'Alembertian, it can be seen that the covariant equation (5.21) is a form of the wave equation (5.1) for each component μ, taking $c = 1$ as usual. This is hardly surprising as the propagation of light signals was used as the foundation of special relativity theory.

The source of the field disturbance is the charged current vector j_μ. The discussion of section 5.1 applies equally well to the present case; in particular the solution (5.3) becomes

$$A_\mu(\boldsymbol{r}, t) = \int \left[\frac{j_\mu(\boldsymbol{r}', t')}{|\boldsymbol{r} - \boldsymbol{r}'|}\right]_{ret} d^3\boldsymbol{r}'. \qquad (5.24)$$

It is interesting to actually carry out the integration of (5.24) in the special case of a single point charged particle in arbitrary motion. Substituting from (5.17) into (5.24), the integrals may be performed with a little manipulation[9] to yield

$$A_\mu(x) = \left[\frac{e\dot{z}_\mu}{(x^\nu - z^\nu)\dot{z}_\nu}\right]_{ret} \qquad (5.25)$$

where $[\kappa]_{ret}$ now denotes evaluation of the quantity in brackets on the backward null cone from the point x of interest. In three-vector notation the result (5.25) may be written

$$\phi(x) = \left[\frac{e}{\kappa R}\right]_{ret}$$
$$\boldsymbol{A}(x) = \left[\frac{e\boldsymbol{v}}{\kappa R}\right]_{ret} \qquad (5.26)$$

where $\kappa = 1 - \boldsymbol{n} . \boldsymbol{v}$, $\boldsymbol{n} = \hat{\boldsymbol{R}}$ and $\boldsymbol{R}$ is the instantaneous vector $(\boldsymbol{r} - \boldsymbol{r}')$. The four-vector A_μ has been expressed as $(\phi, \boldsymbol{A})$.

The expressions in (5.26) are known as the Liénard–Wiechert potentials. They may be differentiated to give the fields $\boldsymbol{E}$ and $\boldsymbol{B}$; for example

$$\boldsymbol{E}(\boldsymbol{r}, t) = e\left[\frac{(\boldsymbol{n} - \boldsymbol{v})(1 - v^2)}{\kappa^3 R^2}\right]_{ret} + e\left[\frac{\boldsymbol{n}}{\kappa^3 R} \times \{(\boldsymbol{n} - \boldsymbol{v}) \times \dot{\boldsymbol{v}}\}\right]_{ret}. \qquad (5.27)$$

Thus $\boldsymbol{E}$ divides naturally into two components; the first, called the near field, is short ranged (R^{-2}) and depends only on the particle velocity $\boldsymbol{v}$. (It is sometimes called the velocity field.) The second, called the far field, is long ranged (R^{-1}) and depends linearly on the acceleration $\dot{\boldsymbol{v}}$. The far field represents energy and momentum which becomes 'detached' from the charged particle and flows away into space as radiation (the far field is sometimes called the radiation field). In the non-relativistic limit the far field reduces to

$$\left[\frac{e\hat{\boldsymbol{e}}\dot{v}}{R} \sin \Theta\right]_{ret} \qquad (5.28)$$

where Θ is the angle between $\boldsymbol{n}$ and $\dot{\boldsymbol{v}}$. The electric polarization unit vector, $\hat{\boldsymbol{e}}$, lies in the plane of $\boldsymbol{n}$ and $\dot{\boldsymbol{v}}$, and is perpendicular to $\boldsymbol{n}$, i.e. the radiation is a transverse wave. The direction of $\hat{e}$ is taken to be negative if it points along the $\dot{v}$ direction.

The rate of flow of energy away into space as a result of the far field may be ascertained from (5.27) by evaluating the Poynting vector $(4\pi)^{-1}\ \boldsymbol{E} \times \boldsymbol{B}$. The result is, in the non-relativistic limit

$$\frac{e^2}{4\pi}\, \dot{v}^2 \sin^2 \Theta \tag{5.29}$$

power per unit solid angle. Note that electromagnetic radiation may only be produced by *accelerating* charged particles.

If a particle radiates energy in the form of electromagnetic waves, it must suffer a damping to its mechanical energy to pay for it. The magnitude of the corresponding damping force may easily be obtained by equating the power radiated (5.29) to the work done by the damping force, f_{rad}, in some periodic motion. After averaging Θ to give a factor $8\pi/3$, this yields, for motion between times t_1 and t_2

$$\int_{t_1}^{t_2} f_{rad} \cdot \boldsymbol{v}\, dt = -\int_{t_1}^{t_2} \tfrac{2}{3}\, e^2 \dot{\boldsymbol{v}} \cdot \dot{\boldsymbol{v}}\, dt = \tfrac{2}{3} \int_{t_1}^{t_2} e^2 \ddot{\boldsymbol{v}} \cdot \boldsymbol{v}\, dt - \tfrac{2}{3} e^2 (\dot{\boldsymbol{v}} \cdot \boldsymbol{v}) \Big|_{t_2}^{t_1}.$$

The second expression has been obtained by an integration by parts. The second term of this expression will vanish for periodic motion, to give

$$\int_{t_1}^{t_2} (f_{rad} - \tfrac{2}{3} e^2 \ddot{\boldsymbol{v}}) \cdot \boldsymbol{v}\, \mathrm{d}t = 0$$

whereupon we identify

$$f_{rad} = \tfrac{2}{3} e^2 \ddot{\boldsymbol{v}}. \tag{5.30}$$

The nature of this damping force is very strange. Unlike other forces in physics, it depends on the *second* derivative of the velocity, i.e. rate of change of acceleration, a fact which will prove of great significance. The reader may wonder what is the origin of this damping force. Thus far, we have only been dealing with fields generated by special types of given source distributions. The picture is not complete, however, unless the effect of the fields back on the sources is included, as given by the Lorentz law of force (5.22). In this equation the nature of the field $F_\nu{}^\mu$ was left unspecified. In fact, this field is the *total* field incident on the ith particle *including the field of the particle itself.* It is this self-action which allows a charged particle to radiate energy in the absence of all other charges or fields in this theory.

Dirac[10] has given a derivation of the radiation reaction, or damping force f_{rad} by evaluating the self-action term covariantly in the following way (except where necessary the tensor indices will henceforth be omitted for simplicity of writing)

$$F_{tot} = F_{ret} + F_{in} \tag{5.31}$$

$$= F_{adv} + F_{out} \tag{5.32}$$

where the total field, F_{tot}, is either decomposed into the 'incident' field from outside the volume V, F_{in}, plus the retarded contribution of the charges inside V, F_{ret}, or alternatively into the outgoing field F_{out}, and the advanced contributions from the sources. Dirac defined a *radiation field* F_{rad} as

$$F_{rad} = F_{out} - F_{in} \tag{5.33}$$

which, because of (5.31) and (5.32) may be written

$$F_{rad} = F_{ret} - F_{adv}. \tag{5.34}$$

Therefore F_{rad} agrees with the earlier definition (5.28) of the radiation field at large distances from the charged particle and at large times after the acceleration has taken place, because F_{adv} is zero there. The present definition has some advantages though. The corresponding potential $A_{ret} - A_{adv}$ is a solution of the homogeneous equation

$$\square^2 A_\mu = 0 \tag{5.35}$$

which follows from the definition (5.33). Solutions of (5.35) have the property that if they vanish on the surface S at all times (see figure 5.2) they vanish everywhere (this follows from (5.6) with $\rho = 0$). Consequently, $F_{rad} = 0$ everywhere when the particle acceleration is zero, and not just at future null infinity.

It is helpful to introduce other fields $\bar{F}$ and $\bar{A}$, defined by

$$\bar{F} = \tfrac{1}{2}(F_{ret} + F_{adv}) \tag{5.36}$$

$$\bar{A} = \tfrac{1}{2}(A_{ret} + A_{adv}) \tag{5.37}$$

which are time symmetric. These fields are tied to sources, i.e. satisfy the inhomogeneous equations. For example

$$\square^2 \bar{A}^\mu = j^\mu. \tag{5.38}$$

As remarked in section 5.1, appropriate solutions of the homogeneous equation can always be added to these fields to obtain solutions which satisfy the particular boundary conditions of the problem. For example

$$F_{ret} = \bar{F} + \tfrac{1}{2}F_{rad} \tag{5.39}$$

$$F_{adv} = \bar{F} - \tfrac{1}{2}F_{rad}. \tag{5.40}$$

The significance of the fields F_{rad} and $\bar{F}$ for the radiation damping force turns out to be crucial. Rewriting (5.31) as follows:

$$F_{tot} = \bar{F} + \tfrac{1}{2}F_{rad} + F_{in} \tag{5.41}$$

Dirac proceeded to evaluate the effect of F_{tot} when the surface S of figure 5.2 surrounds a single charged particle. The first two terms in (5.41) represent this particle's self-fields, while the third term represents the incoming field due to all the other particles in the world outside V (and any radiation from past infinity). The calculation is rather lengthy; the result shows that the $\bar{F}$ field, acting on its own source particle with charge e, contributes to the rest energy of the particle an amount of order e^2/a, where a is a characteristic dimension of the size of the source particle. This rest energy *diverges* for a point source, $a \to 0$. Such divergences have plagued the subject of both classical and quantum mechanics for many years. Their removal is at present only a formal device, called renormalization, based on the fact that observable quantities always turn out to be the finite difference between two diverging quantities. The infinities are therefore merely subtracted out and ignored.

The F_{rad} field does not create any divergence problems. It is finite everywhere, even on the world line of the source particle. It in fact gives rise to the self-force of the particle, as described by the covariant equation

$$f^\mu = \tfrac{2}{3}e^2(\dddot{z}^\mu + \ddot{z}^\nu \ddot{z}_\nu \dot{z}^\mu). \tag{5.42}$$

In the non-relativistic limit f_{rad} is recovered from f^μ.

Finally, the F_{in} term of equation (5.41) leads to the usual Lorentz force $\dot{z}^\nu F^\mu_{in\,\nu}$. The equation of motion of a (renormalized) charged particle is therefore

$$m\ddot{z}^\mu - \tfrac{2}{3}e^2\, \dddot{z}^\mu + \ddot{z}^\nu \ddot{z}_\nu \dot{z}^\mu = e\dot{z}^\nu F^\mu_{in\,\nu}. \tag{5.43}$$

5.4 Preacceleration

The non-relativistic form of (5.43) may be written

$$m(\dot{v} - \epsilon \ddot{v}) = f_{ext} \tag{5.44}$$

where $\epsilon = \dfrac{2}{3}\dfrac{e^2}{m}$, and is known as the relaxation time of the particle. It is usually very small; for an electron, about 10^{-23} seconds, or 10^{-13} cm in our units. f_{ext} is the total external force due to the right-hand side term of equation (5.43), together with any force of non-electromagnetic origin. If $f_{ext} = 0$, equation (5.44) has *two* solutions

$$\dot{v} = 0 \tag{5.45}$$

and

$$\dot{v} = \dot{v}(0)e^{t/\epsilon}. \tag{5.46}$$

The second solution (5.46), known as the 'runaway' solution, is clearly unphysical because it represents an unlimited self-acceleration, which would render all charged particles unstable if allowed. The existence of this strange solution can be traced to the nature of the damping force, f_{rad}, which depends on the second derivative of the velocity (equation (5.30)). To see this, write the general solution of (5.44) in the following way

$$\dot{v} = e^{t/\epsilon}\left[\dot{v}(0) - \frac{\epsilon^{-1}}{m}\int_0^t e^{-t'/\epsilon} f_{ext}(t')\, dt'\right] \tag{5.47}$$

where a particular integral of (5.44) has been added to the complementary function (5.46). In a normal mechanical problem, which involves a second order in time differential equation, it is sufficient to specify the position and velocity of a particle at one instant in order to arrive at a unique solution. In the present case (5.44) is *third* order in time, so that it is necessary to specify also the initial acceleration to uniquely determine the trajectory.

Now an arbitrary initial acceleration $\dot{v}(0)$ will lead to runaway solutions in (5.47), but Dirac[10] proposed that these be eliminated by fixing a special value of the initial acceleration instead of leaving it arbitrary. Inspection of (5.47) shows that if we choose

$$\dot{v}(0) = \frac{1}{m}\int_0^\infty e^{-t'/\epsilon} f_{ext}(t')\, dt' \tag{5.48}$$

then v remains finite as $t \to \infty$. Choosing $\dot{v}(0)$ by (5.48) enables (5.47) to be rewritten as

$$\ddot{v} = \frac{1}{m}\int_0^\infty e^{-s} f_{ext}(t + \epsilon s)\, ds. \tag{5.49}$$

Because the particle now obeys an integrodifferential equation of motion (5.49), its behaviour at any time t depends upon the external forces acting on the particle at *later* times than t. In particular, the initial acceleration (5.48) depends not on the instantaneous force at $t = 0$, but on a weighted average over the entire future motion of the particle. For example, suppose that the particle suffers a δ function force in one dimension at $t = 0$. Putting $f_{ext}(t) = K\,\delta(t)$ in equation (5.49) yields

$$\begin{aligned} \dot{v}(t) &= \frac{K}{m}\, e^{t/\epsilon} \qquad && t < 0 \\ &= 0 && t > 0 \end{aligned} \tag{5.50}$$

so that the particle actually accelerates before the force is applied (see figure 5.4). This phenomenon is called *preacceleration*, and appears to violate our usual notions of 'causality'. However,the characteristic time scale of such acausal behaviour is ϵ, which in practice is far too small to be observable; it is of the same order as the time required for light to cross the classical 'radius' of an elementary particle, and quantum considerations become important well before that.

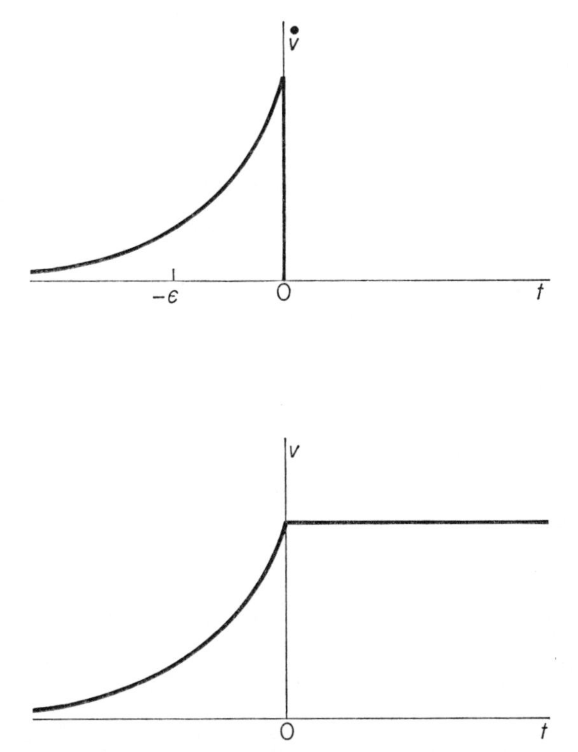

Figure 5.4 Preacceleration from an impulse. The charged particle receives a sharp blow at $t = 0$, but it must accelerate *before* this moment to avoid runaway motion.

Somehow (if equation (5.44) can be taken seriously) a feeling of dissatisfaction remains. The runaway solutions are eliminated at the expense of introducing unpleasant acausal behaviour, which in turn is disregarded because of the values that the constants of nature (e and m) happen to have. Unless it can one day be shown that nature is constrained in some fundamental way to choose values of the constants to preserve causality, this explanation will not be totally convincing.

5.5 Boundary conditions for radiation

In section 5.3 it was discussed how the retarded field F_{ret} of a single charge, when acting on itself, could be profitably decomposed into a source free part $\frac{1}{2}(F_{ret} - F_{adv})$ which gives rise to the observable, finite radiation damping force, and a part with a source on the particle $\frac{1}{2}(F_{ret} + F_{adv})$ which leads to an indistinguishable part of the self-energy (mass m) of the particle. Nothing has yet been said about F_{in}.

Consider a collection of charged particles in a volume V bounded by a surface S in an otherwise empty world. The total force acting on particle i is therefore due to the field

$$\sum_{j \neq i} F_{(j)ret} + \tfrac{1}{2}(F_{(i)ret} - F_{(i)adv}) + F_{in} \tag{5.51}$$

but now F_{in} includes only the source free fields coming from infinity, as there are supposed to be no charged particles outside the volume V (see section 5.1).

However, there is another possibility. Nowhere in the derivation of (5.51) have any boundary conditions been used, so that the theory must still preserve its basic symmetry in time, consistent with the Maxwell–Lorentz equations. This is indeed so, for we could have started with equation (5.32) instead, in which case the following expression would have resulted in place of (5.51)

$$\sum_{j \neq i} F_{(j)adv} - \tfrac{1}{2}(F_{(i)ret} - F_{(i)adv}) + F_{out}. \tag{5.52}$$

We could even take the average of (5.51) and (5.52) to obtain a manifestly time symmetric form

$$\tfrac{1}{2}\sum_{j \neq i} (F_{(j)ret} + F_{(j)adv}) + \tfrac{1}{2}(F_{in} + F_{out}). \tag{5.53}$$

If we wish to obtain the usual retarded fields that are experienced in the real world it is necessary to impose a boundary condition on the system

$$F_{in} = 0 \tag{5.54}$$

which is known as the Sommerfeld radiation condition. Inspection of (5.51) shows that the fields acting on particle i are only the *retarded* fields of the other particles, plus the self-field (finite part). Moreover, from (5.33) and (5.34) it is found that the total outgoing field is now

$$F_{out} = \sum_{\text{all } j} (F_{(j)ret} - F_{(j)adv})$$

so that (5.52) reduces to (5.51)—the time symmetry has been destroyed.

Of course, if instead the condition $F_{out} = 0$ is imposed, a description with advanced fields from the other particles would be obtained, together with radiation which converges onto charged particles to accelerate them with an 'anti-damping' force which has the opposite sign to the usual damping force.

In the case of a system of charges permanently localized in an otherwise empty Minkowski space, there is no property of the charges or the space which can supply a temporal asymmetry, and there is no way in which F_{in} and F_{out} could be distinguished. The only meaningful boundary conditions in such a world would be

$$F_{in} = F_{out} = 0 \tag{5.55}$$

in which case it follows from (5.53) that the field acting on a charged particle i is

$$\tfrac{1}{2}\sum_{j\neq i}(F_{(j)ret}+F_{(j)adv}). \tag{5.56}$$

5.6 Cosmology again

In the real world we do not have a permanently localized distribution of charged particles in an otherwise empty Minkowski space. Matter is assumed to be distributed throughout Riemannian space, and the space–time manifold may well possess asymmetric properties. For example, it may be forever expanding (($\dot{R}/R > 0$), or have a singularity in one time direction only; there may also be horizons of one sort or another. In addition to the properties of the space, the particles themselves may possess temporally asymmetric properties through thermodynamics (or K mesons). All of these properties may have relevance to the choice of boundary conditions in electrodynamics.

One could perhaps justify putting $F_{in} = 0$ in the Friedmann models on the basis that the backward null cone does not extend to past infinity anyway, but intersects a singularity. In the region of this singularity matter and radiation are strongly coupled, with the matter density becoming infinite; under these circumstances the notion of a source free asymptotic field is obviously of dubious validity. (A further complication is that the spacelike surface S may not remain spacelike as the singularity is approached[11].)

Another possible justification for assuming the Sommerfeld boundary condition (5.54) in some cosmological models has been suggested by Sciama[4], who has reasoned that if F_{in} were not zero, it may be infinite. This would be because the cumulative effect of radiation from all 'future' sources might diverge, for the same type of reason as Olbers' paradox (section 4.5). The energy density of the incoming radiation is readily obtained by integrating equation (4.20) over the future

$$\varepsilon_\gamma(t) = R^{-4}(t)\int_t^\infty L(\tau)\rho(\tau)R^4(\tau)\,d\tau. \tag{5.57}$$

(compare (4.21)). Assuming that the number of sources is conserved, $\varepsilon_\gamma(t)$ will diverge for any t if

$$\int_t^\infty L(\tau)R(\tau)\,d\tau \to \infty$$

which demands that $L(t)$ fall off slower than

$$L(t) \propto [R(t)t]^{-1}. \tag{5.58}$$

For example, in the Einstein–de Sitter universe $R(t) \propto t^{\frac{2}{3}}$, so that if the sources cool slower than $t^{-\frac{5}{3}}$ the energy density diverges.

Actually, it's not that simple. If it is accepted that the temporal direction of entropy increase is in the direction of increasing R, then we must allow for the fact that absorption of this radiation will occur, especially in the early fireball. (If this is not so we should be more inclined to say that retarded radiation in a contracting universe is being discussed.) This absorption will, for example, occur by pair creation, but eventually the created pairs will annihilate again, and in so doing radiate electromagnetic energy. There will be an F_{in} field associated with this process also, and this F_{in} field will in turn be responsible for creating more pairs, and so on. Indeed, the only energy to be removed from this escalation is that lost from the cosmological expansion. It therefore seems that there is an inbuilt instability in any coexistence of entropy increase and advanced radiation processes.

It is unlikely that anyone will be very impressed by the sort of reasoning employed in this section. Most students of electrodynamics would rather accept the boundary condition (5.54) as a fact of life, and ignore cosmological complications. There is, however, one case in which the vanishing of F_{in} can always be justified, because it does not conflict with the time symmetry of the system. This is the case when F_{out} vanishes also, i.e. boundary conditions (5.55). This possibility will now be explored.

5.7 The absorber theory of radiation

Return for the moment to the simple case of a permanently localized collection of charged particles in an otherwise empty Minkowski space. Inspection of equation (5.53) indicates that the overall time symmetry of the electromagnetic fields is preserved if the following boundary conditions are chosen

$$F_{in} + F_{out} = 0. \tag{5.59}$$

Thus, every charged particle generates retarded and advanced fields symmetrically. The total field of the whole collection acting on the ith charged particle is given by expression (5.56). It is obvious that a system of particles with boundary condition (5.59) will behave very differently from a system for which the usual Sommerfeld condition is imposed. However, there is one case in which the results obtained by using (5.59) and (5.54) will coincide. This is the case that $F_{out} = 0$ anyway, so that (5.59) implies (5.54). Such a situation will occur if all the charged particles are confined inside an *adiabactic enclosure*. In section 3.1 it was discussed how such an enclosure, in which the contents are isolated from the outside world, cannot possess an asymmetry in time on thermodynamic grounds. This fact emphasizes once again the close connection between the electrodynamic and thermodynamic asymmetries. Of course, when such an enclosure is in a state of thermodynamic equilibrium all radiative processes are exactly balanced by their inverses, so that the notion of advanced and retarded radiation is not really applicable. Any fluctuation from equilibrium would produce both advanced and retarded radiation.

The meaning of this statement can be understood in the context of quantum electrodynamics. Consider the well-defined problem of an excited atom inside a perfectly reflecting box. There is a definite probability after a certain time that the atom will be in its ground state and a photon emitted. However, there is a small but finite probability per unit time that the photon, being confined to a finite volume by the box, will be reabsorbed. There is no problem here about retarded or advanced potentials; of course, we may perhaps wish to refer to the first (emission) process as the production of retarded radiation and the second (absorption) process as the removal of retarded radiation, but the basic time symmetry is clearly present when the corpuscular quality of radiation provided by quantum theory is employed. The description here is of none other than a Poincaré cycle for quantum systems (see section 6.2). To pass from a quantum to a classical description it would be necessary to consider a very large number of atoms emitting a cloud of photons in a coherent wave. The recurrence time would then be very much longer; nevertheless, the eventual formation of contracting coherent waves is assured. This is not really surprising in view of the discussion of the 'pond' problem in section 5.2, for we are now dealing with *finite* systems again. If the box were opened the photon would escape, never to return (if space were otherwise empty), just as removing the stopper from a flask of gas would cause the eternal dissipation of the gas molecules into the empty space around.

Infinite systems can still behave like bounded adiabatic enclosures if they are *completely absorbing* or opaque to electromagnetic radiation; that is, if a sufficiently large volume can be found which behaves like a light-tight box. In the idealized case of a localized distribution of charged particles in an otherwise empty Minkowski space, this means that the absorptive properties of the outer layers are so great that radiation emitted by particles near the centre cannot penetrate to the empty space region outside the system. This would imply $F_{out} = 0$, and hence, with boundary conditions (5.59), $F_{in} = 0$ also. Inspection of (5.51), (5.52) and (5.53) shows that all the following expressions for the total field acting on a particle i near the centre of such an opaque distribution are equal

$$\tfrac{1}{2}\sum_{j\neq i}(F_{(j)ret} + F_{(j)adv}) \tag{5.60}$$

$$\sum_{j\neq i} F_{(j)ret} + \tfrac{1}{2}(F_{(i)ret} - F_{(i)adv}) \tag{5.61}$$

$$\sum_{j\neq i} F_{(j)adv} - \tfrac{1}{2}(F_{(i)ret} - F_{(i)adv}). \tag{5.62}$$

It is instructive to see in detail how these expressions can be equal for a simplified model of the absorber. The model treated here is the one given by Wheeler and Feynman in their original paper on the theory of time symmetric electrodynamics[12]. Consider a particle i with charge e located at the centre of a spherical cavity whose radius is large enough for the interior cavity wall, of radius R_{int}, to be in the radiation zone of i. The wall consists of free charged particles labelled by an index j. The density of these particles is ρ, and they each have mass m and charge e. The cavity wall has unspecified depth, but its outer radius, R_{ext}, is supposed to be great enough for the wall to be opaque.

First consider the form (5.60) for the fields. Each particle gives rise to $\frac{1}{2}$ advanced + $\frac{1}{2}$ retarded far fields when accelerated. Suppose that i is given a non-relativistic acceleration for a short time. It will radiate electromagnetic waves into the past and future symmetrically according to the expression being used, and these waves will strike the cavity walls, thereby setting other charges into motion. This motion will in turn produce time symmetric fields from the walls. Because there is clearly a complicated network of interactions between past and future motions, it is not possible to reason through a causal chain of response and, for example, ascertain the outcome of an arbitrary motion of i. Instead, an attempt can only be made to find self-consistent solutions for the whole assembly of charges. Specifically, it would be expected from the result (5.61) or (5.62) that a collective motion of the system might be found in which each particle produces just the right fields so that the superposition of these fields from every particle is equivalent to using the fully retarded formulation (5.61), or the fully advanced (5.62).

Therefore, the following cycle of reasoning will be now used. The particle i is set into motion, as a consequence of which a fully retarded field is produced as the end result. This field travels out across the cavity until it strikes the inside of the cavity wall, and starts to penetrate it. The wall particles will be set into motion by this field, and this motion will give rise to $\frac{1}{2}$ retarded + $\frac{1}{2}$ advanced fields. Consider a typical particle j; the retarded component will travel on into the wall as usual, but the advanced component can be pictured schematically as a spherical wavefront collapsing onto j from across the cavity (see figure 5.5). Because the cavity is large, this advanced disturbance appears as a very nearly plane wave moving from the left in the figure, sweeping across the particle i at the moment of its acceleration, and passing on to converge eventually onto j at the same moment as the retarded wave from i sets it into motion.

The reader will have noticed that the advanced 'response' wave from j travels across the space between i and j with the retarded wave from i, so these two waves would be expected to interfere. Wheeler and Feynman demonstrated that this interference is constructive, and sufficiently great to just enhance the $\frac{1}{2}$ retarded wave from i to the full retarded wave assumed

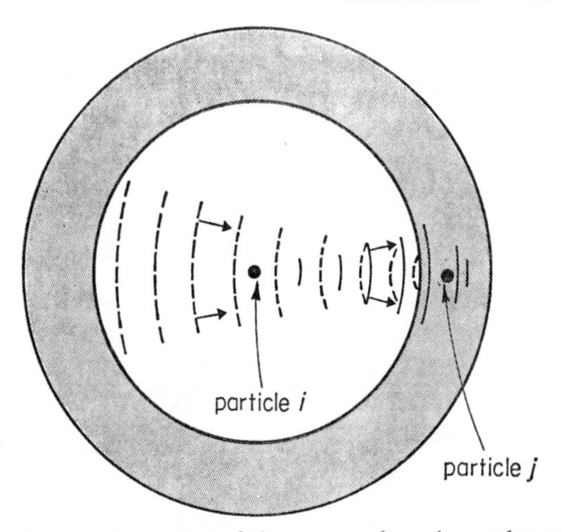

Figure 5.5 Particle i, at the centre of the opaque box, is accelerated and emits a retarded wave (unbroken lines) in the direction of particle j, which is set into motion as a result. The advanced wave due to this motion of j sweeps across the cavity from the left (broken lines), crossing the centre at the moment that i is accelerated, and then accompanying the retarded wave of i, eventually to converge on j at the moment that it is set into motion.

originally. In addition to this, there will also be interference to the left of i in figure 5.5, but this time with the advanced wave from i. This interference is destructive and just cancels the $\frac{1}{2}$ advanced wave from i. The solution described is therefore self-consistent; the original retarded wave from i is actually found to be $\frac{1}{2}$ retarded from i plus $\frac{1}{2}$ advanced from j.

The nature of the interference may appear a little puzzling at first sight. In the region of i the advanced field of j is nearly a plane wave, whereas the fields of i are spherical, shrinking to a point at the centre of the cavity. Furthermore, the response field of the distant particle j will be much weaker in this region than the fields of i. However, we have only discussed the motion of one of the wall particles j. The argument presented above clearly does not depend on the distance between i and j, so that the particles behind and in front of j will respond similarly, and their advanced fields will combine together near i. The same reasoning may be used for all the wall particles at all orientations from the centre. The total effect will be the superposition of a large number of nearly plane waves, all moving in slightly different directions; this appears as a spherical wavefront imploding onto the centre of the cavity (see figure 5.6), passing through particle i, and exploding outwards again towards the walls. Thus, both the primary fields of i, and the *combined* response fields of the cavity wall are spherical in shape, as required for their mutual interference.

It remains to show that the response field of the cavity wall has the right

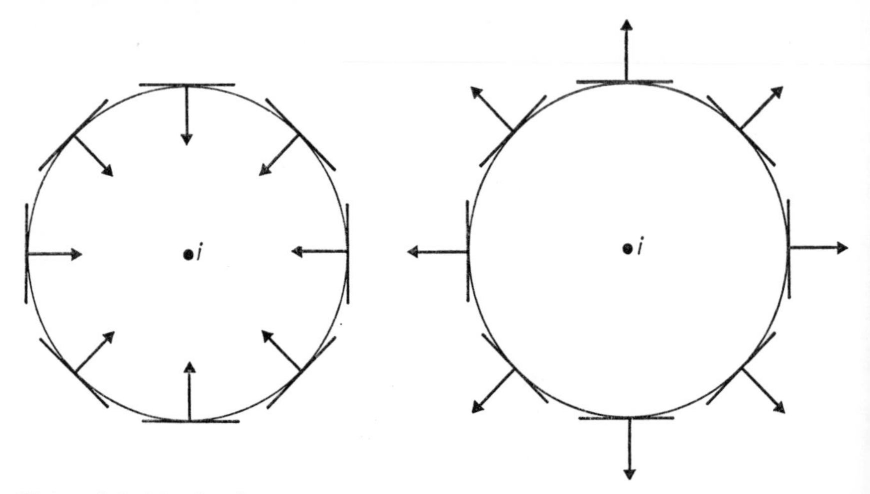

Figure 5.6 Nearly plane response waves from all orientations when superposed appear as a spherical wavefront converging on the particle i at the centre, passing through, and emerging as a diverging spherical wave. The shape of the total response wave is therefore the same as the waves from i, with which it interferes.

amplitude and phase to combine with the $\frac{1}{2}$ retarded + $\frac{1}{2}$ advanced field of i and produce the fully retarded field assumed originally. The proof given by Wheeler and Feynman[12] will now be outlined.

Suppose that i is set into motion non-electromagnetically, resulting in a non-relativistic acceleration given by

$$\dot{\boldsymbol{v}} = \boldsymbol{U}_0 e^{-i\omega t}. \tag{5.63}$$

A more general acceleration can always be built up from a Fourier superposition with different frequencies ω, because the theory is linear. $\boldsymbol{U}_0$ is a constant amplitude vector. The electric component of the far field at j produced by this acceleration is obtained from equation (5.28)

$$e\hat{\boldsymbol{e}}R^{-1} \sin \Theta \,.\, U_0 e^{-i\omega(t-R)}$$

where R is the interparticle separation. The acceleration of j due to this electric field will be

$$e^2(mR)^{-1}\hat{\boldsymbol{e}} \sin \Theta \,.\, U_0 e^{-i\omega(t-R)}. \tag{5.64}$$

We wish to obtain the $\frac{1}{2}$ advanced response field at the cavity centre generated by this acceleration of j. Expression (5.28) may be used again, except that $[\ ]_{ret}$ is replaced by $[\ ]_{adv}$ (although the remarks which follow (5.28) concerning the vector directions still apply). First it is noted that the acceleration of j is in the direction of $\hat{\boldsymbol{e}}$, and so perpendicular to $\boldsymbol{n}$, the unit vector along the radius from i to j (see figure 5.7). The associated electric far field is thus antiparallel to $\hat{e}$ (allowing for propagation delay). At the cavity centre this field is therefore in the plane of $\boldsymbol{U}_0$ and $\boldsymbol{n}$, with a component along the *positive* direction of $\boldsymbol{U}_0$. The magnitude of this component is

$$\tfrac{1}{2}e^3(mR^2)^{-1}\sin^2\Theta\,.\,U_0e^{-i\omega t}. \tag{5.65}$$

The field given by expression (5.65) has two important properties: (1) it depends on R^{-2}, (2) there is no retardation factor in the exponent—the field acts simultaneously with the motion of i.

It is easy to evaluate this response field also in the region around the centre, by adding an appropriate phase factor to (5.65) (there is no need to worry about the amplitude; this varies only slightly across the central region as the wave from j is nearly plane there). In figure 5.7 the various vectors and angles used in the calculation have been drawn. $\boldsymbol{e}$, $\boldsymbol{n}$ and $\boldsymbol{U}_0$ are all in the same plane, but $\overrightarrow{OP}$ is not if we wish to evaluate the field at a general point P. The angular displacement of $\overrightarrow{OP}$ from $\boldsymbol{n}$ and $\boldsymbol{U}_0$ is denoted by θ and ϕ respectively. It will be readily seen that the appropriate phase factor for evaluating the field at P is $\exp(i\omega r\cos\Theta)$, with $OP \equiv r$.

There will actually be another phase factor when the dispersive effect that would always be present in the wall medium is taken into account. If the refractive index of this material is $\varkappa$ there will be a phase lag in the primary retarded wave from i of $-\omega(\varkappa - 1)(R - R_{int})$ when it has penetrated to a depth $R - R_{int}$ through the wall medium to reach particle j.

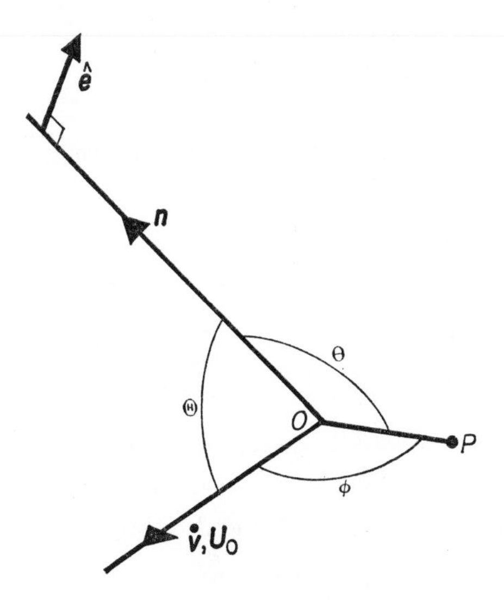

Figure 5.7 The various angles and vectors used in the absorber theory calculation.

To evaluate the *total* $\frac{1}{2}$ advanced electric field component parallel to U_0 at the point P, equation (5.65) must be multiplied by the density of particles ρ and an elementary volume $R^2\,dR\,d\Omega$, then integrated over the volume of the cavity wall. This integral will therefore be, remembering the phase factors

$$\tfrac{1}{2}eU_0e^{-i\omega t}\int e^{i\omega r\cos\Theta}\sin^2\Theta\,d\Omega\int_{R_{int}}^{R_{ext}}\rho e^2m^{-1}e^{-i\omega(n-1)(R-R_{int})}\,dR. \quad (5.66)$$

The total component perpendicular to U_0 will vanish by symmetry. The radial integral may be dealt with first. Nothing has yet been said about the all important absorptive properties of the cavity wall. This may be taken into account by making n a complex quantity:

$$n = n_0 - ik \quad (5.67)$$

where the small imaginary part represents some as yet unspecified absorption mechanism. Even though k is small, it is assumed that the wall is thick enough to be completely opaque. Thus R_{ext} may be replaced by ∞ without error as the upper limit to the radial integral. This integral is then immediately performed, to give

$$\rho e^2m^{-1}[-i\omega(n-1)]^{-1}. \quad (5.68)$$

Expression (5.68) simplifies if the well-known formula for the refractive index of a dilute plasma of free-charged particles is used (see section 5.9)

$$n = 1 - \frac{2\pi\rho e^2}{m\omega^2} \quad (5.69)$$

which reduces (5.68) to $-i\omega/2\pi$.

The angular integral in (5.66) may be performed using standard geometrical techniques. It has the value

$$\frac{8\pi}{3}\left\{\frac{\sin\omega r}{\omega r}+(\tfrac{3}{2}\sin^2\phi-1)\left[\frac{\sin\omega r}{\omega r}-\frac{3\sin\omega r}{\omega^3r^3}+\frac{3\cos\omega r}{\omega^2r^2}\right]\right\}. \quad (5.70)$$

If P is many wavelengths from 0 (but still with $r \ll R_{int}$), the last two terms in (5.70) may be neglected and the following obtained

$$\frac{4\pi\sin\omega r}{\omega r}\sin^2\phi \qquad (\omega r \gg 1). \quad (5.71)$$

The final expression for the $\frac{1}{2}$ advanced response field at P is from (5.66)

$$-\frac{eU_0}{2r}\sin^2\phi\{e^{i\omega(r-t)} - e^{-i\omega(r+t)}\}. \quad (5.72)$$

To see that this is the desired result, it is recalled from expression (5.28) that the component of the retarded electric far field parallel to $\boldsymbol{U}_0$ is negative, and equal to

$$\left[-\frac{e\dot{v}\sin^2\Theta}{r}\right]_{ret} \tag{5.73}$$

which in the case under consideration is

$$-\frac{eU_0\sin^2\Theta}{r}e^{i\omega(r-t)}. \tag{5.74}$$

The corresponding advanced field is

$$-\frac{eU_0\sin^2\Theta}{r}e^{-i\omega(r+t)}. \tag{5.75}$$

Inspection of figure 5.7 shows that, due to the axial symmetry of the fields about $\boldsymbol{U}_0$, Θ and ϕ measure the same angle as far as the field strength is concerned. Consequently expression (5.72) has the value

$\frac{1}{2}$ retarded field of i $-$ $\frac{1}{2}$ advanced field of i.

It is also of interest to evaluate the advanced response field at the position of i. In this case the opposite limit for (5.70) must be used, i.e. $\omega r \ll 1$. This gives the value $8\pi/3$ for the angular factor as the term in square brackets becomes negligible in this limit. Using this result in (5.66), we obtain for the component of the electric field along $\boldsymbol{U}_0$

$$\tfrac{2}{3}eU_0(-i\omega)e^{-i\omega t} = \tfrac{2}{3}e\ddot{v}. \tag{5.76}$$

Now this electric field will interact with i as it plunges through. The corresponding force along the $\dot{\boldsymbol{v}}$ direction is $\frac{2}{3}e^2\ddot{v}$, which is just the usual force of radiative damping, f_{rad}, given by equation (5.30).

These remarkable results may now be summarized. Starting with the time symmetric fields as given by expression (5.60), and no radiative damping term, we have superposed the collective fields of a model opaque enclosure and demonstrated how the end result is the same as if we had used the fully retarded fields of the particles with the usual radiative damping term, i.e. expression (5.61). This is possible because the advanced response field of the cavity wall converts the time symmetric field of the radiating particle to a fully retarded field, and at the same time interacts with this particle to produce the required force of radiative damping.

It may seem surprising that these results are independent of the precise structure of the cavity wall. The crucial reason for this is the cancellation which occurs in the radial integral between $(n - 1)$ and $\rho e^2/m$. This cancellation is not fortuitous. If a more sophisticated model for the absorbing medium is adopted (including a quantum mechanical treatment[13]) the cancellation still occurs, because the same factors will always appear in the refractive index as appear in the field radiated by j. This is not really surprising as they are calculated the same way. In their original paper Wheeler and Feynman[12] also included the effects of reflection from the interior cavity wall and the interaction between neighbouring absorbing particles, as well as recovering the relativistic radiation damping force (5.42).

5.8 Action-at-a-distance electrodynamics

Although the discussion of the last section is interesting in its own right, it is still based on the apparently arbitrary assumption that (5.59) is the correct boundary condition for the fields. It would be much more satisfactory if a theory of electrodynamics could be developed in which this arbitrariness was removed, and the time symmetry of the fields derived rather than postulated. One obvious way in which this can be achieved is by removing the electromagnetic field *altogether*, for then all solutions of the homogeneous equation, such as F_{in} and F_{out}, would necessarily be zero. Such a theory would reproduce the results of Maxwell electrodynamics in an adiabatic enclosure where $F_{in} = F_{out} = 0$ anyway. There is in fact an additional motivation for removing the independent mechanical degrees of freedom of the field. The self-energy divergences mentioned in section 5.3 arise from the interaction of a particle with its own electromagnetic field; a removal of the field removes the divergences with it.

Before a field-free electrodynamics is discussed, it is helpful to understand the origin of the field concept in physics in the first place. In Newtonian mechanics, the motion of a system of interacting point particles with positions $\boldsymbol{r}_i$ and masses m_i may often be characterized by a Lagrangian L of the form

$$L = \sum_i \tfrac{1}{2} m_i \left(\frac{\mathrm{d}\boldsymbol{r}_i}{\mathrm{d}t}\right)^2 - V(|\boldsymbol{r}_i - \boldsymbol{r}_j|). \tag{5.77}$$

V is an arbitrary two-particle potential, assumed to be a function only of the interparticle separation. The Euler–Lagrange equations are

$$\frac{d}{dt}\left(\frac{\partial L}{\partial \dot{\boldsymbol{r}}_i}\right) - \frac{\partial L}{\partial \boldsymbol{r}_i} = 0 \tag{5.78}$$

which for the Lagrangian (5.77) yield a set of second order differential equations for the motion of each particle i:

$$m_i \frac{d^2\boldsymbol{r}_i}{dt^2} = -\frac{\partial V}{\partial r_{ij}} \hat{\boldsymbol{r}}_{ij} \tag{5.79}$$

where $\boldsymbol{r}_{ij} = \boldsymbol{r}_i - \boldsymbol{r}_j$.

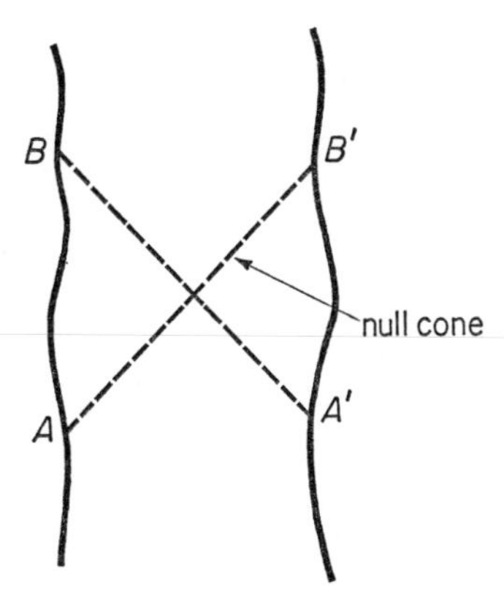

Figure 5.8 The portions of the world lines between AB and $A'B'$ are separated by a spacelike interval, and may be separately specified.

The simplicity of this description is due mainly to the existence of planes of absolute simultaneity (section 1.2), which enables the t variables used in (5.77)–(5.79) to be chosen the same for all particles i. Action and reaction are instantaneous in Newtonian theory, and the independence of L under time translations implies conservation of energy. The set of equations (5.79) may be integrated to give a unique solution if $\boldsymbol{r}_i$ and $\dot{\boldsymbol{r}}_i$ are specified at one instant.

This straightforward treatment breaks down in relativistic mechanics for three reasons: (1) the interparticle potential $V(|\boldsymbol{r}_i - \boldsymbol{r}_j|)$ is not Lorentz invariant, (2) there is no common particle time t, (3) influences between particles can only propagate at finite velocity; action and reaction are not simultaneous. In addition to these difficulties it does not appear that an initial value problem exists in relativistic particle mechanics for the following reason. In figure 5.8 two world lines and the null cone are drawn. The portions of the world lines between AB and $A'B'$ cannot influence each other directly, and may be chosen arbitrarily. A specification of just the position and velocity of each particle at one time is not sufficient to determine the dynamical motion, the entire portions AB, $A'B'$ being required.

The traditional way of circumventing these problems is to introduce into the system an infinite number of additional degrees of freedom which have the effect of 'keeping account' of what the world lines are doing in the portions AB, $A'B'$. This extra mechanical system is called a *field*. It enables the construction of a relativistic field/particle mechanics, such as Maxwell–Lorentz theory based on the action integral (5.20). Instead of particles interacting non-locally across a spacelike interval, the system couples through *point* particle–field interactions. The finite propagation time for influences to

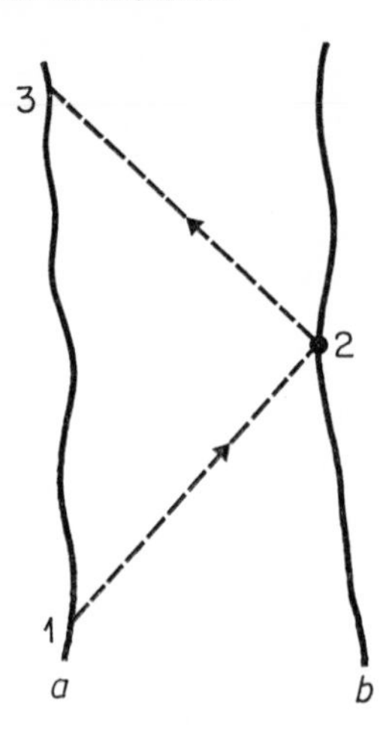

Figure 5.9 Retarded action of two particles *a* and *b*.

pass between particles is accommodated by allowing the field to carry the energy and momentum. Action and reaction balance each other at every particle–field interaction. In addition to this, the introduction of a field leads to new types of phenomena; for example, a particle interacting with its own field produces the emission of energy in the form of radiation.

There still exists the possibility of a relativistic action-at-a-distance mechanics, but with the planes of simultaneity of Newtonian particle mechanics replaced by the null cone of special relativity as the surface of interparticle action. One example of this Lorentz invariant coupling is depicted in figure 5.9, where a particle *a* is acting directly on another particle *b* along the future null cone $1 \rightarrow 2$ (retarded action). Actually, this type of interaction is unacceptable because *b* cannot react back on *a* along $2 \rightarrow 1$, but would instead react on *a* at 3, which may be made arbitrarily later than 1 by increasing the separation of the particles. An acceptable arrangement is shown in figure 5.10, where *both* forward and backward null cones are employed symmetrically. With such a time symmetric interaction particle *a* acts on *b* both at 2 and 3, and *b* can react back on *a* in the same way. It is then possible to satisfy the usual conservation laws in an asymptotic sense[14].

Mathematically, such an interaction may be expressed using the Dirac δ function. The null cone between the two world lines $z_{(i)}$, $z_{(j)}$ satisfies the equation

$$(z^{\mu}_{(i)} - z^{\mu}_{(j)})(z_{(i)\mu} - z_{(j)\mu}) = 0. \tag{5.80}$$

The left-hand side of (5.80) will be abbreviated to $(z_i - z_j)^2$. Then the quantity $\delta[(z_i - z_j)^2]$ vanishes everywhere except where the backward and forward null cones from one intersects the world line of the other. Writing out the null interval (5.80) explicitly yields

$$(z_i - z_j)^2 = (\mathbf{z}_{(i)} - \mathbf{z}_{(j)})^2 - (z^0_{(i)} - z^0_{(j)})^2 \tag{5.81}$$

and using the identity

$$\delta(x^2 - y^2) = \frac{1}{2x}[\delta(x - y) + \delta(x + y)]$$

we obtain

$$\delta[(z_i - z_j)^2] = \frac{1}{2}\frac{\delta[|\mathbf{z}_{(i)} - \mathbf{z}_{(j)}| - (z^0_{(i)} - z^0_{(j)})]}{|\mathbf{z}_{(i)} - \mathbf{z}_{(j)}|} + \frac{1}{2}\frac{\delta[|\mathbf{z}_{(i)} - \mathbf{z}_{(j)}| + (z^0_{(i)} - z^0_{(j)})]}{|\mathbf{z}_{(i)} - \mathbf{z}_{(j)}|}. \tag{5.82}$$

The first term of (5.82) is one half the retarded kernal of the integral (5.2) (with $c = 1$), while the second term is one half the advanced counterpart. This suggests that a time symmetric direct interparticle action theory of electrodynamics can be achieved from a Langrangian of the following type (for a particle i)

$$L_i = -m_i(\dot{z}^{\mu}_{(i)}\dot{z}_{(i)\mu})^{\frac{1}{2}} - e_i \sum_{j \neq i} e_j \int \dot{z}^{\mu}_{(i)}\, \delta[(z_i - z_j)^2]\dot{z}_{(j)\mu}\, d\lambda_j. \tag{5.83}$$

The first term of (5.83) is the usual kinetic energy term for the free particle i, while the second term replaces the potential energy V of the Newtonian theory. This term includes the effect of the interaction of the charge e_i of

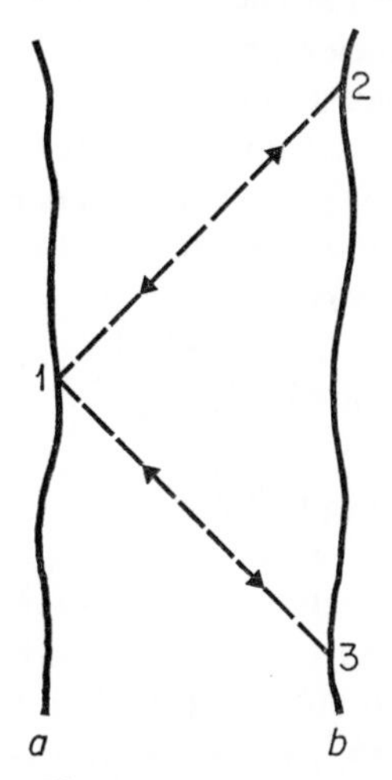

Figure 5.10 Time symmetric action of a and b.

particle i, via the time symmetric direct coupling, with all the other charges e_j on the particles $j \neq i$. Notice that this Lagrangian is non-local in time, containing as it does an integral over all the other particle world lines. As a result, it leads to integrodifferential equations of motion rather than the usual differential equations.

The dynamical behaviour of the system may be recovered from a variational principle using the action[15]

$$\mathscr{S} = -\sum_i m_i \int \left(\dot{z}^{\mu}_{(i)}\dot{z}_{(i)\mu}\right)^{\frac{1}{2}} d\lambda_i - \frac{1}{2}\sum_{i \neq j}\sum e_i e_j \iint \dot{z}^{\mu}_{(i)}\, \delta[(z_i - z_j)^2]\dot{z}_{(j)\mu}\, d\lambda_i\, d\lambda_j \tag{5.84}$$

which should be compared with the conventional particle–field action (5.20). The first term is the same in both cases, merely representing the inertial properties of the particles. However, (5.84) does not contain a term representing the free electromagnetic field action, because the independent field degrees of freedom have been removed in this prescription. The third term of (5.20) is similar to the last term of (5.84) if we define for the latter the quantity

$$A_{\mu}(x) = \sum_j e_j \int \delta[(x - z_j)^2]\dot{z}_{(j)\mu}\, d\lambda_j. \tag{5.85}$$

When used in the action-at-a-distance theory, A_{μ} is understood as a *derived*, auxiliary quantity. Although the language of fields may still be used, it must always be appreciated that they cannot now be chosen independently, but are always determined by the particle motions. In addition, these fields cannot be varied in the action (as they were in section 5.3) to obtain a field equation of motion, which does not exist in this theory.

The last term of (5.84) is symmetric between i and j, and the past and future. The double summation counts once every pair of particles, but there is no term corresponding to a particle acting on itself, hence no divergent self-energies can occur in the theory.

If we operate on equation (5.85) with the d'Alembertian, and use (5.17) and (5.19), the following is obtained:

$$\Box^2 A_{\mu}(x) = 4\pi \sum_j e_j \int \delta^4(x - z_{(j)})\dot{z}_{(j)\mu}\, d\lambda_j = 4\pi j_{\mu}(x)$$

which is just the electromagnetic wave equation of Maxwell theory. (This from of Maxwell's equations requires the Lorentz condition (5.15) to be satisfied. This condition also holds in the present theory, as may be seen by differentiating equation (5.85).) However, this equation must not be interpreted as an equation of motion for the fields, but merely an identity. There is no meaning to the homogeneous equation $\Box^2 A_\mu = 0$, so arbitrary solutions of such an equation cannot be added to A_μ. There is no longer any freedom of choice to match the boundary conditions on the fields; A_μ is completely determined from (5.85) to be $\frac{1}{2}$ retarded + $\frac{1}{2}$ advanced, as are the corresponding $F_{\mu\nu}$. The fields acting on a particle i are those given by (5.60), as expected.

It is obvious that this action-at-a-distance theory will predict the same results as Maxwell field theory in the case of electrodynamic processes which are symmetric in time (e.g. van der Waal's forces). However, the results will not generally be the same for radiative damping, which is asymmetric in time. Nevertheless, as proved in section 5.7, for a system enclosed inside an opaque box radiation damping may be accounted for by the response of the absorber. The action-at-a-distance theory then has the great advantage that the divergent self-energy is absent. If we wish to believe this remarkable theory, we have the added bonus that the existence of fully retarded radiation in the real world implies a condition on the world, namely, that it behaves like the interior of an opaque enclosure. This is a severe constraint on the possible types of cosmology, so that an observation of local radiative phenomena assumes the status of a cosmological observation. On the other hand, one may wish to take the inverse point of view; if conventional cosmological observations do not appear to favour opaqueness, then the action-at-a-distance theory is suspect. In either case it is important to know which cosmological models are transparent and which are opaque. This forms the subject of the next section.

5.9 Is the universe transparent or opaque?

Before we address ourselves directly to the subject of the opacity of the universe, the question of the advanced solutions in the absorber theory must be dealt with. In section 5.7 the equivalence of the two expressions (5.60) and (5.61) was proved. It was also stated that (5.62) was an equivalent form, so argument. This is indeed so, as would be expected from the time symmetry of the fundamental equations. The required cycle of reasoning may be produced by repeating the argument of the retarded case, but in the opposite time sense,
repeating the argument of the retarded case, but in the opposite time sense,
i.e. by assuming a fully advanced primary wave and a $\frac{1}{2}$ retarded response, etc. Of course, there is now an 'anti-damping' force of radiative reaction, in which energy is given up to the accelerated charge by the field. How then are we to explain the observed asymmetry of radiation in the absorber theory after all, if *both* fully advanced and fully retarded solutions are allowed?

Remarkably enough, the answer to this question is tucked away in equation

(5.67). The square root of minus one in front of k can appear with either sign, but the negative sign in (5.67) was chosen so that this term corresponds to *absorption* of radiation, i.e. the fields are exponentially damped away on propagating through the medium, as required by the absorber theory. Now the sign here cannot in fact be chosen at will, but is determined by the processes occurring in the absorbing medium, and nothing has yet been said about how the absorption takes place.

Classically, an electromagnetic wave propagating through an ionized plasma (the model that was taken for the absorbing medium in section 5.7) will set into motion the plasma ions, which in turn will radiate in the forward direction (Thomson scattering), leading to the familiar dispersion of the wave. Not all of the energy imparted to the ion will be re-radiated however. During the forced vibration of the ions their thermal motion may bring them into collision with other ions, with the result that the electromagnetic energy of the wave will be damped and converted into thermal energy of the plasma. This type of process is referred to as *collisional damping* in plasma physics, and *inverse bremmstrahlung* in quantum mechanics. In order that the thermal energy isn't simply re-radiated, it must become spread among many degrees of freedom by multiple collisions. This absorption is clearly an irreversible *thermodynamic* damping effect; the entropy of the absorbing medium increases. This thermodynamic asymmetry in the absorber imposes an asymmetry on the electromagnetic radiation, by permitting the transport of energy from the source at the centre of the cavity to the cavity wall, but not the other way round. The advanced self-consistent solution, which is allowed on purely electrodynamic grounds, is thus ruled out as being overwhelmingly improbable, because it would require the cooperative 'anti-damping' of all the particles in the cavity wall, corresponding to a positive imaginary part in (5.67), i.e. exponentially growing disturbance. Ions would become collisionally excited, and radiate at the precise moment throughout the wall to produce a coherent converging wave to collapse onto the cavity centre at just the moment that the charged particle there was accelerated.

This is an important result. In the absorber theory of radiation the close relationship between electrodynamic and thermodynamic temporal asymmetry is fully exploited. The existence of retarded 'radiation' is assured by the thermodynamic properties of the absorbing medium. The time direction of electromagnetic radiation is determined by the time direction of entropy increase in the universe. This is entirely in accordance with the discussion in section 5.2, and the first two paragraphs of section 5.7. If the system of charges is permanently enclosed in an isolated box, all asymmetries are expected to disappear, whether Maxwell theory or action-at-a-distance is used, and of course in such a system they are equivalent. Fluctuations and Poincaré cycles will occur, with the rare event in which the entropy decreases when the walls of the box emit a collapsing cloud of photons onto a charged particle at the centre. This would be interpreted as 'advanced radiation'. It would occur as frequently as 'retarded radiation'.

It is now necessary to consider which cosmological models are opaque on the future null cone. In the real universe absorption will occur in many ways—by dense objects in galaxies, such as stars, galactic gas and dust, and any intergalactic gas.

Assume that galactic dust, and dense objects such as stars, planets, black holes, etc, absorb their geometric cross section from the photon flux irrespective of frequency ω. The rate of absorption by a given object is then proportional to the photon density ρ_γ, and hence to R^{-3}. Provided that R does not increase faster than $t^{\frac{1}{3}}$, the absorption rate will fall off slower than t^{-1} as the universe expands. The total absorption in the future will therefore diverge logarithmically, i.e. the universe will be opaque. This result may also be derived from considerations of a single photon; the probability that such a photon will be absorbed in a time dt whilst passing through absorbing objects of density ρ and cross section σ is $1 - e^{-\rho\sigma dt} \simeq \rho\sigma\, dt$. This probability is unity if the integral of this latter quantity diverges

$$\int^{\infty} \rho\sigma\, dt = \infty. \tag{5.86}$$

For objects of constant σ, condition (5.86) has the limiting case $\rho \propto t^{-1}$ or $R \propto t^{\frac{1}{3}}$. In the steady-state theory (section 7.2) both ρ and σ are constant, so (5.86) is satisfied in this case also.

Galactic gas (mainly hydrogen) would be an efficient absorber until the galaxy cools sufficiently for ionic recombination to occur. This decouples the gas from the radiation, and the absorption ceases. However, any intergalactic gas may not suffer this fate, owing to the fact that it is being continually expanded by the cosmological motion, which might be sufficiently great to prevent complete recombination (except, of course, in the recontracting models—see p. 146.) According to Bates and Dalgarno[16], the probability per unit time for a given electron to recombine with a proton is proportional to $\rho_i T_i^{-0.75}$, ρ_i and T_i referring to the ion density and temperature respectively. Complete recombination of all hydrogen ions will therefore occur if

$$\int^{\infty} \rho_i T_i^{-0.75}\, dt = \infty \tag{5.87}$$

which requires $\rho_i T_i^{-0.75} \propto t^{-1}$ or slower. In section 7.1 it will be shown that $T_i \propto t^{\frac{2}{3}} R^{-\frac{8}{3}}$, so that all models which expand faster than $t^{\frac{1}{2}}$ will always have some unrecombined ions left. This includes the Einstein–de Sitter ($t^{\frac{2}{3}}$) and open Friedmann ($R \propto t$) models.

These free ions will be capable of absorbing photons by inverse bremmstrahlung down to any frequency. This is important, because the cosmological red shift will eventually reduce any given photon frequency to an arbitrarily low value. In fact very little absorption will in general occur before the photon has travelled a distance of the order of the Hubble radius, so that our discussion may be restricted to the long wavelength (classical) limit $\omega \to 0$. Most absorption by this mechanism is due to collisions of electrons with heavy particles (protons, atoms). The effective mean cross section for photon absorption by a given electron is proportional to[17]

$$\sigma = \frac{A\rho}{T_i^{\frac{1}{2}}\omega^3}(1 - e^{-\omega/kT_i}) \tag{5.88}$$

where ρ now denotes the total heavy particle density, and A contains numerical factors, the constants e, m and k, and a logarithmic factor the slow variation of which may be ignored. The exponential factor in (5.88) may also be neglected because $\omega(\propto R^{-1})$ cannot decrease faster than the ion temperature T_i (recall equations (4.16) and (4.18)).

Making use of the fact that ρ, T_i and ω are proportional to R^{-3}, $t^{\frac{2}{3}}R^{-\frac{8}{3}}$ and R^{-1} respectively, σ in (5.88) is proportional to $t^{-\frac{1}{3}}R^{\frac{4}{3}}$. The electron density also falls like R^{-3}, so that substituting into (5.86) it is found that absorption will be complete if

$$\int^{\infty} R^{-3}t^{-\frac{1}{3}}R^{\frac{4}{3}}\, dt = \infty;$$

the limiting case is

$$R \propto t^{\frac{2}{5}}. \tag{5.89}$$

This result will also be derived thermodynamically in section 7.1.

The conclusion is that, within the framework of conventional matter-conserving cosmological models, complete absorption by the intergalactic medium is impossible. The condition for complete absorption (5.89) is incompatible with the condition (5.87) for the existence of any ions to do the absorbing.

The situation is somewhat different for the recontracting models that return to a high density state in the future. Transforming condition (5.86) to an integral over R we find

$$\int^{0} \frac{\sigma}{R^3\dot{R}}\, dR = \infty. \tag{5.90}$$

As the final singularity is approached $R \to 0$ and $\omega \to \infty$. The dominant absorption mechanism under these circumstances is pair production, for which the cross section σ tends to a constant value as $\omega \to \infty$. The integral (5.90) therefore diverges, indicating that this model is opaque if R collapses like $t^{\frac{1}{3}}$ or faster into the final singularity. This is certainly so in general relativity, where $R \propto t^{\frac{1}{2}}$ in this region (see section 4.2).

A number of authors[18–21] have attempted to derive these results from classical formulae involving the refractive index of an ionized plasma. These attempts may appear somewhat obscure to the casual reader, as they involve lengthy discussions on conformal transformations, space curvature and horizons, none of which are relevant to the results. In addition all these attempts led to wrong answers. To set the record straight, and to tie in with the discussion of section 5.7, the relevant aspects of the theory of refractive index will be reviewed*.

The integrand $\rho\sigma$ of (5.86) is called the *absorption coefficient*, and is related to κ, the imaginary part of the refractive index n, by the equation[22]

$$\rho\sigma = 2\kappa\omega. \tag{5.91}$$

Suppose equation (5.88) is rewritten in the following notation

$$\sigma = \frac{\omega_p^2 \nu}{\omega^2 \rho} \tag{5.92}$$

where ω_p is called the *plasma frequency* and is equal to $(4\pi\rho\, e^2/m)^{\frac{1}{2}}$. ν is the effective frequency of collision of a given electron with heavy particles

$$\nu = \left(\frac{4\pi e^2}{m}\right)\frac{A\rho}{T_i^{\frac{3}{2}}\omega}\,(1 - e^{-\omega/kTi}). \tag{5.93}$$

In addition to inverse bremmstrahlung, photons will also undergo Compton scattering, which in the long wavelength limit reduces to classical Thomson scattering which has a cross section $\sigma_T = 8\pi\, e^4/3m^2$. This is of course the mechanism which provides the dispersive properties of the medium, and does not represent real absorption. However, it will still remove energy from the primary wave, and so must be included in κ. A distinction is sometimes drawn between the cross section of Thomson scattering and that due to true absorption processes by referring to the latter as the *reaction* cross section[23]. Rewriting σ_T in terms of the electron relaxation time ϵ

$$\sigma_T = \omega_p^2 \epsilon/\rho \tag{5.94}$$

and adding this expression to (5.92) yields the total *effective* cross section. From (5.91), this corresponds to an imaginary part of the refractive index of

$$\kappa = \frac{1}{2}\frac{\omega_p^2}{\omega^2}\left(\omega\epsilon + \frac{\nu}{\omega}\right). \tag{5.95}$$

* The application of electrodynamics to Riemannian space requires some care[11]. However, owing to the conformal invariance of Maxwell's equations, the situation in Robertson–Walker space (which is conformally flat) is equivalent to that in Minkowski space.

Equation (5.95) is normally derived classically from the equation of motion of a charged particle in an oscillating electric field. It is only an approximation valid for $\omega \gg \omega_p$. The general formula is[17]

$$(n_0 - ik)^2 = \left\{1 - \frac{\omega_p^2/\omega^2}{1 + \left[\omega\epsilon + \dfrac{\nu}{\omega}\right]^2}\right\} - i\left\{\frac{(\omega_p^2/\omega^2)\left[\omega\epsilon + \dfrac{\nu}{\omega}\right]}{1 + \left[\omega\epsilon + \dfrac{\nu}{\omega}\right]^2}\right\}. \quad (5.96)$$

In the conventional cosmological models $\dfrac{\omega}{\omega_p} \propto R^{\frac{1}{2}}$, so that $\omega \gg \omega_p$. The square root of the first term in braces on the right-hand side of equation (5.96) may then be expanded to obtain

$$n_0 \simeq 1 - \frac{\frac{1}{2}\omega_p^2/\omega^2}{1 + \left[\omega\epsilon + \dfrac{\nu}{\omega}\right]^2} \quad (5.97)$$

and

$$k \simeq \frac{\frac{1}{2}(\omega_p^2/\omega^2)\left[\omega\epsilon + \dfrac{\nu}{\omega}\right]}{1 + \left[\omega\epsilon + \dfrac{\nu}{\omega}\right]^2}. \quad (5.98)$$

The $\left[\omega\epsilon + \dfrac{\nu}{\omega}\right]^2$ term in the denominators of (5.97) and (5.98) represents the effect of radiative damping (see section 5.4). It is normally completely negligible, and in view of the fact that $\dfrac{\nu}{\omega} \to 0$ (or remains constant) for small ω it will certainly be so in our region of interest. To a good approximation

$$n_0 \simeq 1 - \frac{1}{2}\frac{\omega_p^2}{\omega^2} = 1 - \frac{2\pi\rho e^2}{m\omega^2} \simeq n$$

which was the expression used in equation (5.69). Also in this approximation (5.98) reduces to (5.95).

The reactive part of k (which represents true absorption) is given by the second term in the braces of (5.95). In classical notation, the condition (5.86) for complete opaqueness is

$$\int^{\infty} \frac{\omega_p^2\nu}{\omega^2}\, dt = \infty \quad (5.99)$$

where use has been made of (5.92). From the definition of the collision frequency ν (5.93), it follows that $\nu \propto R^{-\frac{2}{3}}t^{-\frac{1}{3}}$ in this limit, which when substituted into (5.99) yields the previous result $R \propto t^{\frac{2}{5}}$, (5.89).

In his original treatment of the cosmological absorption problem, Hogarth[18] ignored the temperature dependence of the collisional damping, putting $\nu \propto \rho \propto R^{-3}$ in (5.99) and obtaining the erroneous result $R \propto t^{\frac{1}{4}}$ in place of $t^{\frac{2}{5}}$. More recently, Burman[21] has included the temperature dependence, but uses $T_i \propto R^{-2}$ instead of the correct relation $T_i \propto t^{\frac{2}{3}}R^{-\frac{8}{3}}$ (see section 7.1), and also takes the wrong limit $\omega \ll kT_i$ in the exponential $e^{-\omega/kT_i}$, which leads to $\nu =$ constant and $R \propto t$ as the limit for complete absorption.

In the steady-state theory the physical state of the intergalactic medium remains (on the large scale) constant with time due to continual creation of matter. Consequently ρ, ν and ω_p are all constant. One result of this is that $\dfrac{\omega_p}{\omega} \propto R^{-1}$ so that the radiation frequency eventually falls below the plasma frequency. Inspection of (5.96) shows that under these circumstances the real part of n is negative. Physically, this situation corresponds to the fact that the radiation cannot propagate through the plasma, but is reflected. There is an evanescent wave which exponentially decays in the forward direction with skin depth ω_p^{-1} if $\nu = 0$. This does not represent absorption, which must be extracted from (5.96) by taking the square root in the limit $\omega \to 0$. This yields $k \to \omega_p/(2\omega\nu)^{\frac{1}{2}}$ from which it follows that $k\omega \propto R^{-\frac{1}{2}} \propto e^{-\frac{1}{2}Ht}$; the integral (5.86) does not *diverge*.

These results will now be summarized. Complete absorption by intergalactic ionized gas does not occur in any of the conventional cosmological models, or the steady state model. However, there is complete absorption by discrete objects in the steady state model, and the matter conserving models that expand like $t^{\frac{1}{3}}$ (Dirac model) or slower. Finally, there is complete absorption in the final fireball of the oscillating models. Observationally, an expansion rate of $t^{\frac{1}{3}}$ or less would lead to trouble concerning the age of the universe, which would have to be shorter than is generally supposed to be possible. The steady state theory suffers from the acute problem of being unable to account for the cosmic background radiation, and has rather fallen out of favour. On the other hand, the oscillating Friedmann model is broadly in agreement with observations, except in regard to the density measurement (see section 4.2) which many cosmologists would not regard as serious.

To end this chapter, we point out that Wheeler and Feynmann[12] gave another self-consistent solution of the absorber theory, in which there are fully retarded fields acting on any particular charged particle, although the future null cone is actually transparent. To understand how this can be, refer to figure 5.11 which shows an opaque box with a hole cut in the wall, called by these authors 'the passage'. The portion of the wall opposite to the passage is called 'antipassage'. The charged particle i at the centre of the cavity will radiate fully retarded fields to all parts of the box except the passage by the now familiar mechanism of absorber response. In the direction of the passage

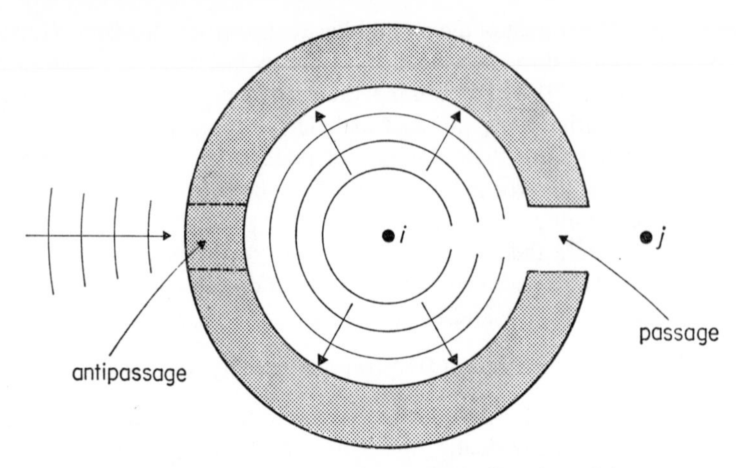

Figure 5.11 Although the wall is transparent in the direction of the passage, a test charge j placed outside the passage will still experience a fully retarded field, and the radiative damping force on i is full strength. The energy which would have been absorbed by the wall in the direction of the passage paradoxically appears instead on the outer surface of the antipassage *before* it is radiated from i.

the fields of i would appear at first sight to be only $\frac{1}{2}$ retarded fields (the $\frac{1}{2}$ advanced field being cancelled by the response field of the antipassage to the retarded field of i in that direction). Moreover, one would expect the radiative damping force to be reduced accordingly. However, the $\frac{1}{2}$ advanced field of i in the direction of the antipassage has not yet been taken into account; in this case it is uncancelled owing to the absense of absorbing material in the passage. In the language of field theory, this $\frac{1}{2}$ advanced field will appear as a disturbance coming in from infinity, and striking the *outer* surface of the antipassage *before* i is set into motion. This field will be absorbed as usual, and in so being will generate response fields. The $\frac{1}{2}$ advanced response from the outside surface will make the incoming $\frac{1}{2}$ advanced field from infinity up to a fully advanced field while the $\frac{1}{2}$ retarded response field will produce the the following effects.

(1) Cancel the $\frac{1}{2}$ advanced field of i through the thickness of the antipassage wall; no disturbance therefore propagates through the wall.

(2) Cancel the $\frac{1}{2}$ advanced field of i inside the cavity, thus removing any advanced effects inside the cavity.

(3) Act on i to make the force of radiative damping up to full strength.

(4) Make the $\frac{1}{2}$ retarded field of i up to fully retarded; a test particle placed outside the passage will actually experience a fully retarded field.

In the direct interparticle action interpretation there can be no question of *energy* propagating out through the passage and away to infinity. The energy emoved from i by the radiative damping force appears (1) on the inside face ›f the box where this is intact, (2) on the outside face of the box opposite vhere it is not intact.

The latter phenomenon is all the more unusual for its occurring *before i* ıas been set into motion. A neat way of depicting this is shown in figure .12, by the two diagrams named by Gold[24] 'zig' and 'zag'. The vertical lines epresent world lines of two particles, one being i and the other a particle in he wall of the box. The heavy portions of these world lines indicate that the article is excited and the wavy line indicates the null cone along which this xcitation energy is passed from one particle to the other. Notice that *both* ig and zag are time symmetric. The usual situation is represented by the zig iagram, with positive energy passing into the future. In the situation dis-ussed above the zag diagram is required to describe the heating up of the uter surface of the antipassage. This appears to be a process in which nega-ive energy is passed along the future null cone from the antipassage to i, but an also be viewed in field theory as the appropriate superposition of positive nergy waves as described earlier.

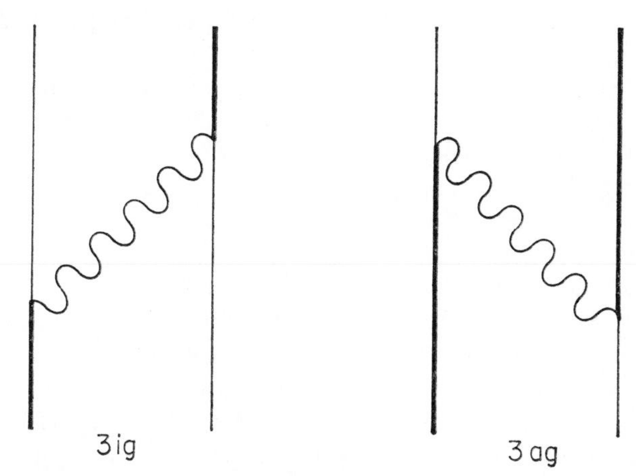

Figure 5.12 Both zig and zag are unchanged by a rotation of 180°.

The reader may find the explicit appearance of advanced effects unsatis-ıctory and prefer to reject this latter type of self-consistent solution. This ıould not be done too lightly for the following reason. In the ever expanding 'riedmann cosmological models, the future null cone is transparent, but long the past null cone the matter is strongly opaque before decoupling. 'hese models therefore possess the properties of figure 5.11. However, any dvanced effects take place only in the remote past, before decoupling, and re quite unobservable. Unfortunately, these models seem to suffer from the scalation effects mentioned in section 5.6. The total energy density output of radiation in the universe has to be absorbed in some way in the earl stages, without heating up the fireball too much. Even if all the matter in th universe was created from this electromagnetic energy, only one photon i 10^8 or so would be accounted for. Nevertheless, the uncertainties surroundin the initial stages do not make this objection compelling.

References

1. See for example W. D. Jackson, *Classical Electrodynamics*, Wiley, Nev York, 1963, section 6.6.
2. P. M. Morse and H. Feshbach, *Methods of Theoretical Physics*, McGraw Hill, New York, 1953, section 6.2.
3. W. D. Jackson, *Classical Electrodynamics*, Wiley, New York 1963 section 6.7.
4. D. W. Sciama, *Proc. Roy. Soc.* A, **273,** 484, 1963.
5. O. Penrose and I. C. Percival, *Proc. Phys. Soc.*, **79,** 605, 1962.
6. K. R. Popper, *Nature*, **177,** 538, 1956; **178,** 382, 1956; **179,** 1297, 1957 **181,** 402, 1958.
7. The spontaneity proviso may be dropped for open systems as pointe out by E. L. Hill and A. Grunbaum, *Nature*, **179,** 1296, 1957.
8. See for example J. L. Anderson, *Principles of Relativity Physics*, Academi Press, New York, 1967, appendix.
9. J. L. Anderson, *Principles of Relativity Physics*, Academic Press, Nev York, 1967, p. 219.
10. P. A. M. Dirac, *Proc. Roy. Soc.* A, **167,** 148, 1938.
11. G. F. R. Ellis and D. W. Sciama, Global and non-global problems i cosmology, in *General Relativity—Papers in Honour of J. L. Syng* (Ed. L. O'Raifeartaigh), Clarendon Press, Oxford, 1972.
12. J. A. Wheeler and R. P. Feynman, *Rev. Mod. Phys.*, **17,** 157, 1945.
13. P. C. W. Davies, PhD thesis, London University, 1970, appendix 1.
14. J. A. Wheeler and R. P. Feynman, *Rev. Mod. Phys.*, **21,** 425, 1949. J. L Anderson, *Principles of Relativity Physics*, Academic Press, New York 1967, sections 7–9.
15. A. D. Fokker, *Z. Phys.*, **58,** 386, 1929.
16. D. R. Bates and A. Dalgarno, chapter 7, in *Atomic and Molecular Pro cesses* (Ed. D. R. Bates), Academic Press, New York, 1962.
17. V. L. Ginzburg, *The Propagation of Electromagnetic Waves in Plasmas* Pergamon Press, Oxford, 1964, p. 488.
18. J. E. Hogarth, *Proc. Roy. Soc.* A, **267,** 365, 1962.
19. F. Hoyle and J. V. Narlikar, *Proc. Roy. Soc.* A, **273,** 1, 1963; **277,** 1 1964.

20. P. E. Roe, *Mon. Not. Roy. Astr. Soc.*, **144,** 219, 1969.
21. R. Burman, *The Observatory*, **90,** 240, 1971; **91,** 141, 1971.
22. See for example R. W. Ditchburn, *Light*, 2nd edition, Blackie, London, 1963, chapter 15.
23. W. D. Jackson, *Classical Electrodynamics*, Wiley, New York, 1963, p. 606.
24. T. Gold, *The Nature of Time*, Cornell University Press, Ithaca, 1967, pp. 35–41.

Further reading

1. D. K. Sen, *Fields and/or Particles*, Ryerson Press, Toronto, 1958.
   J. L. Anderson, *Principles of Relativity Physics*, Academic Press, New York, 1967. Both these books deal extensively with theories of direct interparticle action. The latter also contains a section on preacceleration, and a brief mention of the absorber theory. The absorber theory is also discussed by J. Hogarth, and F. Hoyle and J. V. Narlikar in *The Nature of Time* (Ed. T. Gold), Cornell University Press, Ithaca, 1967, chapters 1 and 2. For a simple account of the absorption properties of cosmological models, and a criticism of earlier work, see P. C. W. Davies, Is the universe transparent of opaque?, *J. Phys.*, A, **5,** 1722, 1972.
2. J. V. Narlikar, Neutrinos and the arrow of time in cosmology, *Proc. Roy. Soc.* A, **270,** 553, 1962.
3. For details of experiments aimed at detecting advanced radiation see R. B. Partridge, Absorber theory of radiation and the future of the universe, *Nature*, **244,** 363, 1973, and M. L. Herron and D. T. Pegg, A possible experiment on absorber theory, *J. Phys.* A (to be published).

6 Time Asymmetry In Quantum Mechanics

6.1 Quantum principle of microreversibility

It is commonly stated that the laws of quantum mechanics are invariant unde time reversal. This statement needs careful attention (the discussion will b restricted to non-relativistic quantum mechanics).

General space–time transformations are represented in quantum theor by linear transformations in Hilbert space of the state vector. If the physica system is to be symmetric under this transformation the observable quant ties must not change under the corresponding Hilbert space transformatio There are two types of observable quantities:

(1) the absolute value of the projection of one state vector $|\phi\rangle$ onto anothe $|\psi\rangle$, i.e. $\langle\psi \mid \phi\rangle$, which gives the probability amplitude that a system repre sented by $|\phi\rangle$ is found on measurement to be in the state $|\psi\rangle$,

(2) the expectation value of an observable β for a state $|\psi\rangle$, $\langle\psi| M |\psi\rangle$. M a linear Hermitean operator whose eigenvectors $|\phi_n\rangle$ span the Hilbert spac completely, so that $\sum_n |\phi_n\rangle\langle\phi_n| = 1$. (Hermitean conjugation will be denote by †.)

If the transformation in the Hilbert space changes $|\phi\rangle \to |\phi'\rangle$, $|\psi\rangle \to |\psi'$ $M \to M'$, we require

$$|\langle\psi' \mid \phi'\rangle| = |\langle\psi \mid \phi\rangle| \quad (6.$$

and

$$\langle\psi'| M' |\psi'\rangle = \langle\psi| M |\psi\rangle. \quad (6.$$

Denote the linear operator which effects the transformation by U. Then

$$|\phi'\rangle = U |\phi\rangle \quad (6.$$

$$U(a |\phi\rangle + b |\psi\rangle) = aU |\phi\rangle + bU |\psi\rangle. \quad (6.$$

Denote the inverse of U by U^{-1}, so that

$$U^{-1} |\phi'\rangle = |\phi\rangle.$$

Now (1) will be invariant under the transformation U if $\langle\psi' \mid \phi'\rangle = \langle\psi \mid \phi\rangle$, e.

$$\langle\psi U^{\dagger} \mid U\phi\rangle = \langle\psi \mid \phi\rangle.$$

'his will be so if $U^{\dagger}U = 1$; that is, if U is unitary, $U^{-1} = U^{\dagger}$. The trans-ɔrmation relations for the operators are obtained with the help of equation 5.2):

$$\begin{aligned}\langle\psi'| M' |\psi'\rangle &= \langle\psi U^{\dagger}| M' |U\psi\rangle \\ &= \langle\psi| U^{\dagger}M'U |\psi\rangle \\ &= \langle\psi| M |\psi\rangle\end{aligned}$$

hich requires

$$U^{\dagger}M'U = M$$

r

$$U^{-1}M'U = M \tag{6.5}$$

this case.

Unitary operators are found to describe all symmetry operations of interest, xcept for time reversal. To see this, consider the unitary operator T which ıanges

$$t \to -t$$

$$\boldsymbol{p} \to -\boldsymbol{p}$$

$$\boldsymbol{r} \to \boldsymbol{r}$$

$$\boldsymbol{\sigma} \to -\boldsymbol{\sigma}$$

here $\boldsymbol{p}$, $\boldsymbol{r}$ and $\boldsymbol{\sigma}$ are the operators of linear momentum, position and spin ıgular momentum respectively. All known physical systems, with the ex-ɩption of those involving K mesons (see section 6.4), possess Hamiltonian ɔerators H that are invariant under T transformations:

$$T^{-1}HT = H. \tag{6.6}$$

The time development of the state vector $|\psi\rangle$ is given by the Schrödinger ɩuation

$$H|\psi\rangle = i\frac{\partial}{\partial t}|\psi\rangle. \tag{6.7}$$

perating on (6.7) from the left with T

$$TH|\psi\rangle = Ti\frac{\partial}{\partial t}|\psi\rangle = -i\frac{\partial}{\partial t}(T|\psi\rangle).$$

owever, it follows from (6.6) that $TH|\psi\rangle = H(T|\psi\rangle)$, so that

$$H(T|\psi\rangle) = -i\frac{\partial}{\partial t}(T|\psi\rangle)$$

or

$$H\,|\psi'\rangle = -i\frac{\partial}{\partial t}\,|\psi'\rangle. \tag{6.8}$$

A comparison of (6.7) and (6.8) shows that $|\psi'\rangle$ does not obey the same equation as $|\psi\rangle$. This is not really surprising, because H is second order in the space variables in non-relativistic quantum mechanics, but equation (6.7) is first order in time. Such an equation would not be expected to display time reversal symmetry.

Nevertheless, a solution of the Schrödinger equation is not itself observable and the symmetry is restored by simply reversing the sign of i in (6.8). This may be effected by defining a new operator $\Theta = TK$, where K is an operator which carries out a complex conjugation on all numbers written after it. Operating on equation (6.7) from the left with Θ

$$\begin{aligned}\Theta H\,|\psi\rangle &= \Theta i\frac{\partial}{\partial t}\,|\psi\rangle \\ &= -i\Theta\frac{\partial}{\partial t}\,|\psi\rangle \\ &= i\frac{\partial}{\partial t}(\Theta\,|\psi\rangle).\end{aligned} \tag{6.9}$$

If H is real $[K, H] = 0$. It then follows from (6.6) that $[\Theta, H] = 0$. Consequently, equation (6.9) yields

$$H(\Theta\,|\psi\rangle) = i\frac{\partial}{\partial t}(\Theta\,|\psi\rangle).$$

Θ is actually a non-linear operator; returning to equation (6.4), we obtain

$$\Theta(a\,|\phi\rangle + b\,|\psi\rangle) = a^*\Theta\,|\phi\rangle + b^*\Theta\,|\psi\rangle$$

and

$$\langle\psi' \mid \phi'\rangle = \langle\psi\Theta^\dagger \mid \Theta\phi\rangle = \langle\psi \mid \phi\rangle^* = \langle\phi \mid \psi\rangle,$$

where * denotes complex conjugation. However

$$|\langle\psi' \mid \phi'\rangle| = |\langle\phi \mid \psi\rangle| = |\langle\psi \mid \phi\rangle|$$

so that Θ still leaves the physical content of quantum mechanics unchanged. Θ is called an *antiunitary* operator.

These considerations are usually summarized in the *quantum principle of microreversibility*. Suppose that we prepare a system in a state $|\psi\rangle$ at a time t_1 and ask for the probability w that it is in the state $|\phi\rangle$ at a later time t_2. The state vector at time t_2 may be obtained in terms of $|\psi\rangle$ by a formal integration of the Schrödinger equation (6.7). This new vector will clearly be

$$U(t_2, t_1)\,|\psi\rangle$$

where

$$U(t_2, t_1) = e^{-iH(t_2-t_1)} \tag{6.10}$$

and is called the *evolution operator*. The probability w will be

$$w = \langle\phi|\, U(t_2, t_1)\, |\psi\rangle. \tag{6.11}$$

Now consider the reverse probability, w_{rev}, that a system at time t_1 in a state $K\,|\phi\rangle$ is found at time t_2 in a state $K\,|\psi\rangle$. This will be

$$w_{rev} = \langle\psi K^\dagger|\, U(t_2, t_1)\, |K\phi\rangle.$$

It follows immediately from the definition (6.10) and the reality of H that

$$U(t_2, t_1) = K^\dagger U^\dagger(t_2, t_1) K$$

whence the principle of microreversibility

$$w = w_{rev} \tag{6.12}$$

for all $|\phi\rangle$, $|\psi\rangle$, t_1, t_2. Symbolically (6.12) may be written

$$\text{prob}(|\psi\rangle \to |\phi\rangle = \text{prob}(K\,|\phi\rangle \to K\,|\psi\rangle)$$

which should be compared with the classical principle (2.14).

In spite of the reversibility of the physical content of quantum mechanics, there are many well-known quantum processes which appear to us asymmetric in time; for example, the radioactive decay of a nucleus. The asymmetry is related to the special initial conditions and the openness of the system considered. To make this quite clear it is helpful to consider a model finite system in which reversibility is retained throughout, and see how the asymmetry enters when the limit of an infinite (open) system is taken.

Consider a complete orthonormal set of eigenstates $|\phi_n\rangle$ in terms of which the general state vector may be expanded

$$|\psi\rangle = \sum_n a_n\, |\phi_n\rangle \tag{6.13}$$

where

$$a_n = \langle\phi_n \mid \psi\rangle$$

and

$$\sum_n |a_n|^2 = 1.$$

For simplicity, systems with degenerate or continuous eigenstates will not be considered. The set of time dependent a_ns may be used to describe the evolution of the system in place of the state vector $|\psi\rangle$. The Schrödinger equation is replaced by the set of equations

$$\dot{a}_n = -i \sum_m H_{nm} a_m \tag{6.14}$$

where H_{nm} are the matrix elements of the Hamiltonian operator

$$H_{nm} = \langle\phi_n| H |\phi_m\rangle. \tag{6.15}$$

The simple model to be considered consists of a collection of systems, each of which can reside only in one of two stationary states called the ground state and the excited state. The ground states all have energy zero, while the energy of the excited state of the nth system is denoted by E_n. At time $t = 0$ system p is prepared in the excited state with energy E_p, while all the other systems are in their ground states. System p is then coupled to the others via an interaction Hamiltonian H_{int}, which for simplicity is assumed to be the same for all the systems. After a time t there is a definite probability that system p will be in its ground state, and one of the other coupled systems excited instead. Let a_m be the probability amplitude of the stationary state $|\phi_m\rangle$ of the total system in which the mth system is excited and all others are in their ground states. This amplitude will be given from equations (6.14) as

$$\dot{a}_p = -i \sum_m H_{int} a_m e^{-i(E_m - E_p)t} \tag{6.16}$$

and

$$\dot{a}_m = -iH_{int}^{\dagger} a_p e^{i(E_m - E_p)t}. \tag{6.17}$$

The set of coupled differential equations (6.16) and (6.17) may be readily solved using the method of Laplace transforms. Substituting for a_m from (6.17) into (6.16) leads to the integrodifferential equation

$$\dot{a}_p(t) = -\sum_m |H_{int}|^2 \int_0^t a_p(\tau) e^{-i(E_m - E_p)(t-\tau)}\, d\tau. \tag{6.18}$$

Taking Laplace transforms of both sides of (6.18), and using the convolution theorem, yields

$$s\mathscr{L}[a_p] - a_p(0) = -|H_{int}|^2 \sum_m \mathscr{L}[a_p]\mathscr{L}[e^{-i(E_m - E_p)t}]. \tag{6.19}$$

Rearranging (6.19), evaluating the last transform, and using the initial condition $a_p(0) = 1$, we obtain

$$\mathscr{L}[a_p] = \left\{ s + |H_{int}|^2 \sum_m \frac{1}{s + i(E_m - E_p)} \right\}^{-1}. \tag{6.20}$$

Nothing has yet been said about the eigenenergies E_m. It will be assumed that all E_m lie in the range

$$E_p + \epsilon < E_m < E_p - \epsilon$$

where

$$H_{int} \ll \epsilon \ll E_p \tag{6.21}$$

(see figure 6.1). Physically, condition (6.21) means that the effect on the transition probabilities of the ground states and excited states with $|E_m - E_p| \gtrsim \epsilon$

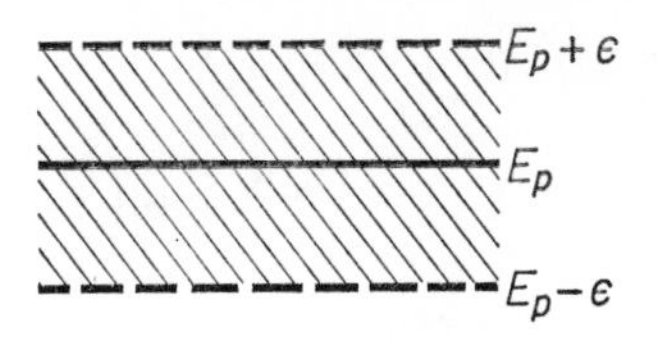

Figure 6.1 The excitation is transferred from system p to one of the other states in the narrow range of energies between $E_p - \epsilon$ and $E_p + \epsilon$.

may be ignored. Furthermore, to simplify matters, we shall choose the levels E_m to be equally spaced by an energy gap ΔE symmetrically about the energy E_p in the range $E_p \pm \epsilon$ (see figure 6.2). Thus

$$E_m - E_p = m\,\Delta E \tag{6.22}$$
$$m = 0, \pm 1, \pm 2, \ldots \pm N.$$

The total number of available upper states is then $2N + 1 \simeq 2\epsilon/\Delta E$. As more and more systems are added, the level spacing ΔE gets smaller, i.e. the density of states $(\Delta E)^{-1}$ gets larger. The summation in (6.20) may be written

$$\sum_{m=-N}^{+N} \frac{1}{s + im\,\Delta E} \tag{6.23}$$

but because of (6.21) the limits $\pm N$ may be extended to $\pm\infty$ without appreciable error. The summation in (6.23) may then be carried out exactly to give

$$\frac{\pi}{\Delta E}\coth\left(\frac{\pi s}{\Delta E}\right). \tag{6.24}$$

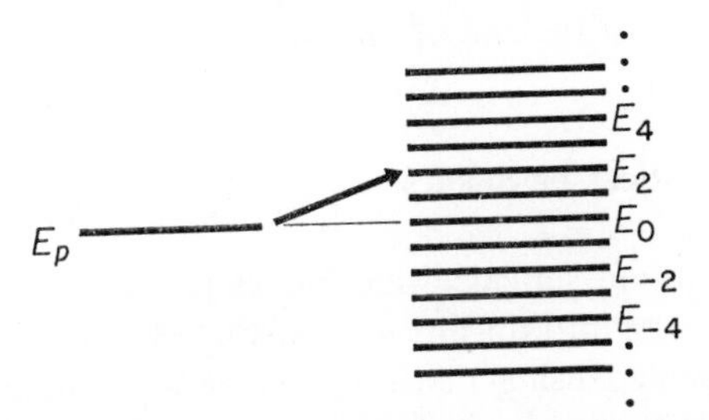

Figure 6.2 The energy range $E_p - \epsilon$ to $E_p + \epsilon$ is divided into $2N + 1$ levels with equal spacing ΔE. As N is increased ΔE is reduced, until a continuum is approached in the limit as $N \to \infty$, corresponding to an open system.

Finally, substituting (6.24) into (6.20) yields

$$\mathscr{L}[a_p] = \left\{ s + \frac{\pi\, |H_{int}|^2}{\Delta E} \coth\left(\frac{\pi s}{\Delta E}\right) \right\}^{-1}. \tag{6.25}$$

Now it may be proved quite generally that for finite N and discrete eigenenergies E_m, the amplitudes a will display Poincaré cycles[1]. Indeed, this may be seen explicitly in the simple case of just two identical coupled systems, the behaviour of which is recovered from (6.25) by taking the limit $\Delta E \to \infty$. In this limit

$$\mathscr{L}[a_p] = \frac{s}{s^2 + |H_{int}|^2}.$$

Inverting the transform, we obtain

$$|a_p|^2 = \cos^2(|H_{int}|\, t) \tag{6.26}$$

which is manifestly symmetric in time, and shows recurrences over a period $2\pi/|H_{int}|$.

As more and more systems are added, the density of states $(\Delta E)^{-1}$ increases, and the Poincaré cycles become longer. The opposite limit $\Delta E \to 0$ corresponds to $N \to \infty$, i.e. an open system. In that case

$$\mathscr{L}[a_p] = \frac{1}{s + \dfrac{\pi}{\Delta E}\, |H_{int}|^2}$$

from which it follows that

$$|a_p|^2 = e^{-2\pi |H_{int}|^2 t/\Delta E}. \tag{6.27}$$

This result is the usual time asymmetric exponential decay of an excited state, with the familiar half life $\Delta E/2\pi\, |H_{int}|^2$. In this limit of an open system, the probability of the excitation returning to p in a Poincaré fluctuation tends to zero. This simple model is a very good description of the photon in the box problem discussed in section 5.7; opening the box corresponds to letting $N \to \infty$. Compare also the classical damping problem treated exactly by Rubin, and described at the end of section 3.5.

6.2 Quantum statistical mechanics

The subject of quantum statistical mechanics is a conjunction of an inherently statistical theory into an already existing statistical framework. This cannot be implemented using phase space, because there now exist finite sized cells of dimensions $\hbar^{6N}$ which represent an *absolute* limitation on our observational resolution. The most elegent formalism in the quantum case is based on the use of the *density matrix*.

The formalism is constructed in the amplitude representation contained in the set of equations (6.14). In terms of these *a*s, the expectation value of an operator M, given by $\langle\psi| M |\psi\rangle$, becomes

$$\begin{aligned}\langle M\rangle &= \sum_n \sum_m a_m^* a_n \langle\phi_m| M |\phi_n\rangle \\ &= \sum_n \sum_m A_{nm} M_{mn} \\ &= Tr(AM)\end{aligned} \tag{6.28}$$

where Tr stands for trace, and A_{nm} is the matrix element $a_m^* a_n$ of a matrix A.

The advantage of using the form (6.28) for $\langle M\rangle$ is that it may be generalized to describe an ensemble of systems. For example, consider a collection of $\mathcal{N}$ systems with possible states $|\psi_i\rangle$, where

$$|\psi_i\rangle = \sum_n a_n^{(i)} |\phi_n\rangle, \qquad i = 1, 2, 3, \ldots \mathcal{N}.$$

Then the *statistical* average $\bar{M}$ of the *quantum* average $\langle M\rangle$ is

$$\begin{aligned}\bar{M} &= \frac{1}{\mathcal{N}} \sum_{i=1}^{\mathcal{N}} \langle\psi_i| M |\psi_i\rangle \\ &= \frac{1}{\mathcal{N}} \sum_{i=1}^{\mathcal{N}} \sum_n \sum_m a_m^{(i)*} a_n^{(i)} M_{mn} \\ &= \sum_n \sum_m \hat{\rho}_{nm} M_{mn} \\ &= Tr(\hat{\rho}M).\end{aligned} \tag{6.29}$$

$\hat{\rho}$ is called the *statistical operator*, or in matrix form, the density matrix. It is defined by

$$\hat{\rho}_{nm} = \frac{1}{\mathcal{N}} \sum_{i=1}^{\mathcal{N}} a_m^{(i)*} a_n^{(i)} = \overline{a_m^* a_n}. \tag{6.30}$$

$\hat{\rho}$ plays a role in quantum statistical mechanics closely analogous to the ρ defined in section 2.4 which represents the density of points in Γ space in the classical theory (for example, compare (6.29) with (2.37)). Moreover, it follows from (6.14) that $\hat{\rho}$ develops in time according to the equation

$$\frac{d\hat{\rho}}{dt} = \frac{\partial\hat{\rho}}{\partial t} - i[\hat{\rho}, H] = 0 \tag{6.31}$$

where the commutator bracket $[\hat{\rho}, H] = \hat{\rho}H - H\hat{\rho}$ is well known to be the quantum analogue of the classical Poisson bracket. Equation (6.31) thus closely resembles Liouville's theorem (2.41).

$\hat{\rho}$ has a number immediate properties

$$Tr(\hat{\rho}) = 1 \tag{6.32}$$

$$0 \leqslant \hat{\rho}_{nn} \leqslant 1 \text{ for all } n \tag{6.33}$$

$$Tr(\hat{\rho}^2) \leqslant Tr(\hat{\rho}). \tag{6.34}$$

In addition, $Tr(\hat{\rho}M)$ is invariant under unitary transformations in Hilbert space. For a transformation matrix S

$$|\phi_p\rangle = \sum_n S_{pn} |\phi_n\rangle.$$

$\hat{\rho}$ transforms like a normal quantum operator

$$\hat{\rho}' = S\hat{\rho}S^{-1}. \tag{6.35}$$

If the exact quantum state of the system of interest is known, all the systems of the ensemble are chosen to be in the same state. The ensemble is then said to be in a *pure state*. In this case $\hat{\rho}_{nm} = \overline{a_m^* a_n} = a_m^* a_n$, so that $\hat{\rho}$ reduces to A, and the theory coincides with the usual wave function picture for a single system. For pure states, it follows from the definition (6.30) that

$$\hat{\rho}^2 = \hat{\rho} \tag{6.36}$$

and from (6.32) that

$$Tr(\hat{\rho}^2) = 1 \tag{6.37}$$

whereas for mixtures

$$Tr(\hat{\rho}^2) < 1. \tag{6.38}$$

Now equation (6.31) may be formally integrated as follows

$$\hat{\rho}(t) = e^{-iHt}\hat{\rho}(0)e^{iHt} \tag{6.39}$$

but as e^{iHt} is a unitary operator, equation (6.39) is simply a special case of (6.35) with $S = e^{-iHt}$. As $Tr(\hat{\rho}M)$ is invariant under such transformations we obtain, for $M = 1$

$$Tr\, \hat{\rho}(t) = Tr\hat{\rho}(0). \tag{6.40}$$

Moreover, it can be shown that for any integer n

$$Tr[\hat{\rho}(t)]^n = Tr[\hat{\rho}(0)]^n. \tag{6.41}$$

In particular when $n = 2$, equation (6.41) together with (6.37) and (6.38) show that *a pure state cannot evolve into a mixture*.

In analogy with the classical definition (2.47) a fine-grained quantum H function is defined by

$$\sigma = Tr(\hat{\rho} \log \hat{\rho}) \tag{6.42}$$

and a fine-grained entropy

$$S = -kTr(\hat{\rho} \log \hat{\rho}).$$

Note that the entropy of a pure state is zero. Because log $\hat{\rho}$ can be expanded in a power series in $\hat{\rho}$ it follows from (6.41) that during the natural evolution of an isolated system σ and S are *constant*:

$$\frac{d\sigma}{dt} = 0 \qquad (6.43)$$

which is the same as the classical result (2.49).

The development of quantum statistical mechanics now follows very closely the classical theory outlined in chapters 2 and 3. Once again, our interest centres around an adiabatically closed box with smooth rigid walls, filled with a large number of identical weakly-interacting particles at low density. As the system is finite, the energy spectrum is discrete, but being a macroscopic system, the level spacing is much smaller than can be distinguished by macroscopic means. Consequently, these levels may be grouped together into sets (analogous to the stars of Γ space). We suppose that the set ν contains G_ν eigenvalues, with average energy E and spacing ΔE. There will always be a limit to the quantum mechanical accuracy of an energy measurement, say η, in accordance with the uncertainty principle of Heisenberg. It will be assumed that the system is such that

$$\Delta E \ll \eta. \qquad (6.44)$$

The construction of ensembles to represent real physical systems proceeds in the usual way. For example, there exist stationary ensembles where $\hat{\rho}$ is a function only of the Hamiltonian; in such cases it follows from (6.31) that $\hat{\rho}$ is constant. Particularly important once again are the uniform ensemble

$$\hat{\rho}_{nm} = \hat{\rho}_0\, \delta_{nm} \qquad (6.45)$$
$$\hat{\rho}_0 = \text{constant}$$

which is invariant under unitary transformations in Hilbert space, and the microcanonical ensemble

$$\hat{\rho}_{nm} = \hat{\rho}_0\, \delta_{nm},\ E < E_m < E + \delta E \qquad (6.46)$$
$$= 0 \text{ otherwise}$$

where E_m is the mth level energy eigenvalue. It is supposed that the macroscopic limitation of energy measurement δE is such that

$$\Delta E \ll \eta \ll \delta E. \qquad (6.47)$$

If the energy interval δE contains N non-degenerate energy levels, we choose $\rho_0 = 1/N$.

It is necessary to make a postulate in quantum statistical mechanics analogous to the postulate D of equal *a priori* probabilities in the classical theory. Once again, the stationary nature of the uniform ensemble (6.45) indicates no quantum preference for one set of quantum states over another, so that it is reasonable to make the following assumption:

in the absence of any information about the system, it is equally likely to be found in any of the possible quantum mechanical states. (F)

However, assumption F is not sufficient to ensure invariance of the representative ensemble under the operations of evolution and unitary transformation in Hilbert space, owing to the possibility of quantum mechanical superposition of states[2]. The above mentioned invariance will be achieved though, if the following postulate of *random phases* is made:

in the absence of any information about the system, the phases of different quantum mechanical states are distributed at random. (G)

The representative ensemble consistent with postulates F and G is the uniform ensemble (6.45). If partial information is available about the system, then the appropriate ensemble is constructed so that the amplitudes of the quantum states consistent with this information are equal, and their phases are distributed at random. For example, a knowledge of the total energy to within a range δE requires the use of the microcanonical ensemble (6.46).

It is now possible to proceed almost line for line in analogy with the classical theory. In particular, one may prove a Poincaré recurrence theorem for the wave function, and develop a quantum ergodic theory[3]. The quantum H theorems do, however, contain an additional feature of interest.

First it is noted that a single system quantum H theorem for a weakly-interacting isolated dilute gas may be proved along the lines of the discussion in section 2.3. A definition of H similar to the classical definition (3.1) is[4]

$$H = \sum_i \left[\alpha g_i \log \frac{g_i}{n_i} - (n_i + \alpha g_i) \log\left(\frac{g_i}{n_i} + \alpha \right) \right] \tag{6.48}$$

where g_i are the numbers of eigenstates of a particle in the group i and n_i are the numbers of particles occupying states in this group (analogous to the cells of phase space). α takes the value $+1$ or -1 depending on whether Bose–Einstein or Fermi–Dirac statistics are used respectively (if $\alpha = 0$ (6.48) reduces to the classical definition (3.1), and corresponds to Boltzmann statistics).

Differentiating equation (6.48), we obtain

$$\frac{dH}{dt} = \sum_i \frac{dn_i}{dt} \log\left(\frac{n_i}{g_i + \alpha n_i} \right). \tag{6.49}$$

As in the classical case $\frac{dn_i}{dt}$ cannot be known without first integrating the equations of motion. Consequently, we obtain instead the average values of these quantities, by considering the transition probabilities between quantum states as a result of collisions, and evaluated in accordance with the fundamental postulates F and G. The standard theory of quantum mechanical scattering supplies the following expressions[4] for the average number of transitions per unit time, Z, between particles from initial groups of states (i, j) to final groups (k, l):

$$\begin{aligned} Z_{ij}^{kl} &= A_{ij;kl} n_i n_j (g_k + \alpha n_k)(g_l + \alpha n_l). \\ Z_{kl}^{ij} &= A_{kl;ij} n_k n_l (g_i + \alpha n_i)(g_j + \alpha n_j). \end{aligned} \tag{6.50}$$

The principle of microreversibility (6.12) demands that the transition probabilities A obey the relation

$$A_{ij;kl} = A_{kl;ij}. \tag{6.51}$$

Equation (6.51), which is the same as the classical relation (3.6), is an *essential step* in the proof of the H theorem. It is a consequence of the fundamental postulates F and G, and the microscopic reversibility of quantum mechanics.

If the results (6.50) and (6.51) are used in equation (6.49), the immediate result is obtained:

$$\frac{dH}{dt} \leqslant 0 \tag{6.52}$$

which is the quantum mechanical single system H theorem. The equality in (6.52) only applies for the Bose–Einstein or Fermi–Dirac distributions (according to whether $\alpha = +1$ or -1), which are the quantum analogues of the Maxwell equilibrium distribution in the classical theory.

The quantum $\bar{H}$ theorem for an ensemble of identical systems will now be discussed in slightly greater detail. The probability of finding a system in the state n is $\hat{\rho}_{nn} = \overline{a_n^* a_n}$, which is called the *fine-grained probability*. If we cannot distinguish macroscopically between a group of G_ν neighbouring states, we shall also be interested in the total probability P_ν of finding a member of the ensemble in that group;

$$\mathrm{P}_\nu = \sum_{n=1}^{G_\nu} \hat{\rho}_{nn}. \tag{6.53}$$

A *course-grained density* (or probability) P_{nn} may now be defined as

$$P_{nn} = \frac{P_\nu}{G_\nu} = \frac{1}{G_\nu} \sum_{n=1}^{G_\nu} \hat{\rho}_{nn} \tag{6.54}$$

for the group of states with common similar properties labelled by n. Both $\hat{\rho}_{nn}$ and P_{nn} are assumed to be normalized to unity

$$Tr\hat{\rho} = Tr\mathrm{P} = 1. \tag{6.55}$$

A course-grained $\bar{H}$ function may now be defined by

$$\bar{H} = Tr(\mathrm{P} \log \mathrm{P}) = \sum_n \mathrm{P}_{nn} \log \mathrm{P}_{nn} \tag{6.56}$$

which should be compared with the fine-grained $\bar{H}$ function σ of (6.42). Unlike σ, however, $\bar{H}$ will *not* remain constant in time for an isolated system. Suppose that at $t = 0$ an observation indicates that the system is in one of a group of states k. A representative ensemble in accordance with F and G may be set up such that

$$\hat{\rho}_{kk}(t = 0) = \mathrm{P}_{kk}(t = 0) \tag{6.57}$$

and

$$\bar{H} = \sum_k \hat{\rho}_{kk}(0) \log \hat{\rho}_{kk}(0). \tag{6.58}$$

At a later time t, the equality (6.57) is no longer valid. The change in $\bar{H}$ over this time is therefore given by

$$\bar{H}(0) - \bar{H}(t) = \sum_k \hat{\rho}_{kk}(0) \log \hat{\rho}_{kk}(0) - \sum_n \mathrm{P}_{nn}(t) \log \mathrm{P}_{nn}(t). \tag{6.59}$$

By definition P_{nn} will be constant over the group of states n, so that the last term of (6.59) may be replaced by $\sum_n \hat{\rho}_{nn}(t) \log \mathrm{P}_{nn}(t)$, to give

$$\bar{H}(0) - \bar{H}(t) = \sum_k \hat{\rho}_{kk}(0) \log \hat{\rho}_{kk}(0) - \sum_n \hat{\rho}_{nn}(t) \log \mathrm{P}_{nn}(t). \tag{6.60}$$

To treat this expression further, it is necessary to find a relationship between $\hat{\rho}_{nn}(t)$ and $\hat{\rho}_{kk}(0)$. This is obtained by integrating Schrödinger's equation in the form (6.14), which is the appropriate form for a density matrix discussion and which follows from the definition (6.30) for $\hat{\rho}$. In terms of the matrix elements of the evolution operator U of equation (6.10), we have the equations

$$\hat{\rho}_{nm}(t) = \sum_k \sum_l U^*_{ml} U_{nk} \overline{a^*_l(0) a_k(0)} \tag{6.61}$$

$$\hat{\rho}_{nn}(t) = \sum_k |U_{nk}|^2 \hat{\rho}_{kk}(0) + \sum_{k \neq l} \sum U^*_{nl} U_{nk} \hat{\rho}_{kl}(0) \tag{6.62}$$

where the first term on the right of equation (6.62) contains the diagonal elements of $\hat{\rho}$ only, while the second term contains the off-diagonal elements. In the case that $\hat{\rho}_{kl}(0)$ is chosen to have random phases in accordance with assumption G, the second term of (6.62) vanishes. Then

$$\hat{\rho}_{nn}(t) = \sum_k |U_{nk}|^2 \hat{\rho}_{kk}(0) \tag{6.63}$$

from which it follows by simple manipulation that

$$\sum_k \hat{\rho}_{kk}(0) \log \hat{\rho}_{kk}(0) \geqslant \sum_n \hat{\rho}_{nn}(t) \log \hat{\rho}_{nn}(t) \tag{6.64}$$

a result known as the *Klein relation*[5].

If the Klein relation is now added to equation (6.60), the following equation is obtained:

$$\bar{H}(0) - \bar{H}(t) \geqslant \sum_n [\hat{\rho}_{nn}(t)\log \hat{\rho}_{nn}(t) - \hat{\rho}_{nn}(t)\log \mathrm{P}_{nn}(t)].$$

Finally, adding to the right-hand side of this equation the zero quantity $\sum_n [\mathrm{P}_{nn}(t) - \hat{\rho}_{nn}(t)]$, we obtain

$$\bar{H}(0) - \bar{H}(t) \geqslant \sum_n [\hat{\rho}_{nn}(t)\log \hat{\rho}_{nn}(t) - \hat{\rho}_{nn}(t)\log \mathrm{P}_{nn}(t) - \hat{\rho}_{nn}(t) + \mathrm{P}_{nn}(t)]. \quad (6.65)$$

The term in square brackets in (6.65) is never negative, so that we may conclude

$$\bar{H}(0) \geqslant \bar{H}(t) \quad (6.66)$$

which is the quantum mechanical generalized $\bar{H}$ theorem for an ensemble.

6.3 Quantum measurement theory

The decrease of $\bar{H}$ as described by (6.66) can be traced to *two* distinct causes. Firstly equation (6.57), which expresses the equality of coarse- and fine-grained probabilities at the initial instant $t = 0$; at a later time the equality is no longer valid: $\hat{\rho}_{kk}(t) \neq \mathrm{P}_{kk}(t)$. $\bar{H}$ therefore decreases for the same reason as in the classical $\bar{H}$ theorem, i.e. the spreading of the fine-grained $\hat{\rho}$ over the coarse-grained stars or groups of states (see section 3.2). Secondly, the Klein relation (6.64) *already contains* an inequality concerning the *fine-grained* matrices $\hat{\rho}_{kk}$. There is no parallel to this relation in classical mechanics; indeed, the classical quantity $\int \rho \log \rho \, dq \, dp$ remains constant in time by (2.49). In fact, the proper quantum counterpart of (2.49) is (6.43), which is not the quantity of interest in the $\bar{H}$ theorem. The quantity σ in (6.43) is equal to $\sum_k \hat{\rho}_{kk} \log \hat{\rho}_{kk}$ only when $\hat{\rho}$ is diagonal, which is just the case of equality in (6.64). It is therefore the presence of off-diagonal terms in the density matrix which leads to a *new type of time asymmetric behaviour* peculiar to quantum mechanics, and which will now be examined carefully.

As mentioned at the beginning of section 6.2, the subject of quantum statistical mechanics contains statistical features at two distinct levels; those that arise, as in classical statistical mechanics, from the assumed limitation of observational accuracy, and those that are inherent in the statistical nature of quantum mechanics itself. It is the latter that is responsible for the new type of asymmetry. Such asymmetry is present, therefore, even in a *single* quantum system, and is indeed very familiar as the spreading of probability over the various possible quantum states with time. There are many examples of this phenomenon; in particular, the spreading of a quantum wave packet[6] (which obeys the same equation as the classical diffusion process).

Now equation (6.43) may be taken as an expression of the fact that the natural evolution of an undisturbed isolated quantum system does not behave asymmetrically in time, as expected from the quantum microreversibility principle of section 6.1. However, the asymmetry contained in the Klein relation does not arise from this 'natural' evolution. Equation (6.64) is valid only if it is assumed that the off-diagonal elements of $\hat{\rho}$ vanish *initially*—random *a priori* phases. (The dynamical evolution of the system destroys the randomness of the phases initially present.) This initial moment, $t = 0$, corresponds to the initial observation of the system with respect to which the appropriate representative ensemble is constructed in accordance with assumptions F and G. It is then also necessary to make a subsequent observation to determine the later value of $\bar{H}$. In classical mechanics, the observation of a system is not considered to have any dynamical consequences, because it may in principle be made with an arbitrarily small and predictable disturbing effect on the system. But in quantum mechanics, any measurement will have a non-negligible and unpredictable effect on the observable conjugate to that being measured; for example, a position measurement of a particle will always unpredictably disturb the particle's momentum. In statistical mechanics we are dealing with a very large number of particles, and for such macroscopic systems this disturbing effect of measurement will have negligible consequences for the *coarse-grained* behaviour. For this reason the classical discussions of the H theorem, time asymmetry, etc, are unaffected by quantum mechanical considerations. Nevertheless, the quantum measurement process does appear to introduce a *microscopic irreversibility*. There is a close analogy between the destruction of microreversibility in the classical system by coupling to the outside world stochastically through the walls of the container (see section 3.5), and the same destruction in a quantum system due to external coupling with the 'measuring apparatus'.

The nature of the quantum irreversibility is easier to understand than its origin. If a quantum system is initially in a pure state ψ, this may be expressed† as a superposition (equation (6.13)) over the eigenstates ϕ_n of the operator corresponding to some observable α:

$$\psi = \sum_n a_n \phi_n \tag{6.67}$$

which yields a density matrix $\hat{\rho}^{(1)}_{nm}$ given by

$$\hat{\rho}^{(1)}_{nm} = a^*_m a_n \qquad \text{(pure case).} \tag{6.68}$$

† Henceforth, we shall often drop the bra and ket labels for simplicity.

After the measurement of α, the system has a probability $|a_n|^2$ of being found in the state ϕ_n. Suppose it is actually in state ϕ_j; then the process of measurement has abruptly converted the state vector from ψ to ϕ_j:

$$\sum_n a_n\phi_n \to \phi_j \tag{6.69}$$

a phenomenon known as the 'collapse of the wave function'. We are not interested here in the irreversible change which occurs when the human experimenter obtains the *knowledge* about the state j into which the system is thrown, for this is no different from the case of a classical observation, e.g. looking to see whether a coin has come down heads or tails. (There is little reason to suppose that the asymmetry associated with this 'entering into the consciousness of the experimenter' is any more than the usual entropy increase of the brain.) To avoid this inessential complication, it is assumed that the measuring apparatus is not 'read' by the experimenter (or anyone else), so that after the measurement has been performed it can only be said that the different states ϕ_n are likely to be realized with relative probabilities $|a_n|^2$. Thus, the new state vector must be described by a *mixture*, with a density matrix $\hat{\rho}^{(2)}$ given by

$$\hat{\rho}^{(2)}_{nm} = |a_m|^2\,\delta_{nm}. \tag{6.70}$$

This means that if a *second* measurement is made, the expectation value of an observable β with operator M would be

$$\begin{aligned} Tr(\hat{\rho}^{(2)}M) &= \sum_n\sum_m |a_m|^2\,\delta_{nm}M_{mn} \\ &= \sum_n |a_n|^2\,M_{nn}. \end{aligned} \tag{6.71}$$

On the other hand, if *no* first measurement had been made, this expectation value would be, instead,

$$\begin{aligned} Tr(\hat{\rho}^{(1)}M) &= \sum_n\sum_m a_m^*a_nM_{mn} \\ &= \sum_n |a_n|^2\,M_{nn} + \sum_{n\neq m}\sum a_m^*a_nM_{mn} \end{aligned} \tag{6.72}$$

which differs from (6.71) by the presence of off-diagonal terms.

The time asymmetry of the quantum measuring process is clearly revealed in a comparison of (6.68) and (6.70) or (6.71) and (6.72). The measuring process has destroyed the off-diagonal elements of $\hat{\rho}$. The entropy $S = -kTr(\hat{\rho}\log\hat{\rho})$ has thus changed from zero initially (this was a pure state) to

$$-k\sum_n |a_n|^2\log|a_n|^2 > 0 \tag{6.73}$$

on account of the removal of these off-diagonal terms in the above summation. Similarly, the expectation value of any other quantity with operator M, which normally contains interference terms between the different superpose states, as expressed in the double summation $n \neq m$ in (6.72), will lose thes terms if a measurement is made, as described in equation (6.71).

The transformation from a pure state (6.68) to a mixture (6.70) is not pos sible by the unitary evolution operator of the isolated system, as alread mentioned in connection with equation (6.41). But such a transformatio has indeed occurred in the measurement process, so that it appears as if th quantum system under observation is capable of changing in two distinc ways: (1) reversibly during the normal evolution of the system as describe by the Schrödinger equation, (2) irreversibly during the measuremen process, which cannot be described by the Schrödinger equation. However this statement is somewhat paradoxical for the following reason. The meas uring apparatus is itself made up of atoms, which ought in principle to obe the laws of quantum mechanics also. If instead of considering only the objec system to be described quantum mechanically, with the apparatus considere as classical (Copenhagen interpretation[7]), the entire system of object plu measuring apparatus is now regarded as one large coupled quantum system the behaviour of which is (in principle) described by an evolution operato for the *whole* system, then this evolution is reversible (type (1) change). What then, has happened to the irreversibility of the measuring process?

An attempt to incorporate the macroscopic measuring apparatus into th quantum description was made by von Neumann[8] in his 'theory of the meas urement process'. Consider a single microscopic object system I, and a macro scopic system II which is to serve as the measuring apparatus. There will be a observable α in I (which we are trying to measure) which possesses a complet orthonormal set of eigenfunctions ϕ_n, and another observable γ in II (e.g the position of a pointer on a scale) which possesses a complete orthonorma set of eigenfunctions Φ_n. Initially ($t = 0$) system I is the pure state ψ an system II is in one of its eigenstates (say Φ_0, corresponding to the zer position of the pointer). ψ may then be expanded in the ϕ_n

$$\psi = \sum_n a_n \phi_n$$

so that the total pure state of system I + II is initially

$$\Phi_0 \sum_n a_n \phi_n. \tag{6.74}$$

The act of measuring α is to couple I and II together for a short time τ, s that they evolve together according to the Schrödinger equation into a fina state with the following properties: (1) the eigenstates ϕ_n are relativel undisturbed by the interaction, (2) the measured value of α can be obtaine by inspection of the eigenstate Φ_n of II, i.e. the position of the pointer, etc It was shown by von Neumann that there exists a class of Hamiltonians H o e total system I + II that realize these requirements. That is to say, for an belonging to this class we have

$$e^{-iHt}\Phi_0 \sum_n a_n\phi_n = \sum_n a_n\phi_n\Phi_n. \tag{6.75}$$

hus, as a result of coupling I and II together, the transition

$$\Phi_0 \sum_n a_n\phi_n \rightarrow \sum_n a_n\phi_n\Phi_n \tag{6.76}$$

is occurred in accordance with the (reversible) unitary development of the tal quantum system. The final state is therefore *still a pure state*. However, ere is now a *correlation* between ϕ_n and Φ_n, which means that an observa- on of II serves as a measurement of I. Moreover, the probabilities $|a_n|^2$ that will be found in the state Φ_n are equal to the same probabilities that I will found in ϕ_n that were used in (6.67) for the usual description of measure- ent, with the apparatus considered as external (and classical) to the system.

If we now wish to make a second measurement of an observable β with erator M considered as an operator in I only, by the rules of quantum echanics it is necessary to trace the expectation value over the states of II[9] he presence of which should make no difference to the value of M obtained I). Because of the correlation in the state $\sum_n a_n\phi_n\Phi_n$ this procedure will iminate all the off-diagonal interference terms ($n \neq m$) in this expectation lue, so that instead of the result (6.72) we obtain (6.71). Therefore, the fect of carrying out the first measurement is to destroy the interference rms in I as postulated in the Copenhagen interpretation.

In spite of all this, the right-hand side of (6.76) is still a pure state. The nsity matrix of the total system I + II will still possess off-diagonal terms, d satisfy equation (6.37). The total entropy S will remain constant—there is irreversibility. It is at this stage that a profound difficulty in the *interpreta- on* of the quantum mechanical formalism is encountered. A quantum stem is described by a state vector in Hilbert space, which in general is a *perposition* of different states, for example, as described by equation (6.67). n the other hand, in the macroscopic world of human beings, such a super- sition of states does not appear to have meaning. It is felt intuitively that e pointer of a piece of apparatus ought to be either in one position *or* the her, but not in some strange way *both*. Of course, the actual positon of e pointer may not be known (because we may not have looked) and in this se we would be obliged to assign certain probabilities to the various possible sitions. However, this is a purely classical obligation related to our sub- ctive view of the world, and is a concept quite distinct from that of a super- sition. In short, the only type of quantity that has meaning in the real world a *mixture* of states with various probabilities, in which interference terms

between different macroscopic alternatives are absent. It follows that som
where during the process of measurement, a transition must occur betwee
a pure state and a mixture, in order that the act of measurement gives
meaningful result. When the measuring apparatus is considered to be extern
to the quantum system, this transition is assumed to occur in the apparatu
But when the apparatus is also a part of the quantum system this explanatio
cannot hold as we have discovered, and the right-hand side of (6.76) appa
ently describes an 'unreal' situation. In view of the fact that the transitio
from pure state to mixture (or from quantum to 'real' description) go
hand in hand with the time asymmetry of the measurement process, it
important to examine some of the attempts to resolve the paradox.

There have been many ingenious attempts to avoid the above difficulties b
subtle reinterpretations of the quantum formalism. It was pointed out b
von Neumann[8] that the interference terms in $\hat{\rho}$ would vanish if the combine
system I + II were in turn measured by a further system III. However, th
interference terms, although removed from I + II, only reappear in III, an
so on. This apparently infinite chain of measurement will only be broke
if a truly 'external' system, not itself subject to the laws of quantum mechanic
interacts with the total quantum system. Such an exterior system might b
interpreted by some as the 'mind' of the experimenter (not the brain) wh
eventually looks at the pointer of the apparatus.

The destruction of the interference terms is thus identified with the collaps
of the wave function. This explanation, in which the origin of the tempor
asymmetry of the quantum measurement process is located in the consciou
ness of the individual, is referred to as 'psycho-physical parallelism', an
has been championed especially by Wigner[10].

Other attempts have been made to break the chain without introducin
such non-physical suppositions into the analysis. In the Everett[11] interpreta
tion, the chain ends with the wave function for the entire universe, whic
evolves deterministically. The probabilistic aspects of quantum mechanic
are then recovered by supposing that the world is continually splitting into *a*
its quantum alternatives, of which a particular observer is only consciou
of one because these many parallel worlds do not interfere. The direction i
time of the splitting determines the temporal asymmetry of the measuremen
process. Of course, this splitting will depend on the type of processes takin
place in the particular 'branch' of the universe under consideration (e.g.
Stern–Gerlach experiment divides the world in two). These processes in tur
depend on the thermodynamic properties of the branch. Some branche
will represent entropy increasing worlds, while others will represent entrop
decreasing worlds (in which Stern–Gerlach experiments 'run backwards'
The situation here is really a quantum mechanical version of Loschmidt'
paradox; the time development of the total wave function of the univers

must be unitary (time symmetric). Nevertheless, nearly all Everett branches will lead to entropy increasing worlds, while a small set will lead to entropy decreasing worlds. A set of measure zero will lead to constant entropy worlds.

In another interpretation, Bohm[12] proposes a deterministic subquantum world of hidden variables, analogous to the hidden microscopic degrees of freedom in thermodynamics. The act of measurement is described by a new dynamical equation, which is manifestly asymmetric under time reversal. The usual results of quantum mechanics emerge when an assumption about equal *a priori* probabilities, similar to that introduced in statistical mechanics, is made.

A somewhat less exotic resolution of the measurement paradox consists in retaining the interference terms in the apparatus wave function, but asserting that they are unimportant, the justification for this being that in a real piece of apparatus (which is any use) the different eigenstates Φ_n must be *macroscopically distinguishable*, for example, the different positions of a pointer on a scale, the different locations of exposed photographic emulsion on a plate. Because the apparatus is a macroscopic object, the time required to establish coherent overlap of the wave functions Φ_n is fantastically large; much larger than the age of the universe[13]. Therefore, the result of a measurement is to transfer the quantum features of the microscopic object system I into the macroscopic apparatus II, where they will lie safely undetected for an enormous duration. Thus the off-diagonal elements of the density matrix may be deleted to an exceedingly good *approximation*, safe in the knowledge that the mutilation will never be detected in practice. However, the quantum measurement irreversibility has clearly disappeared in this interpretation, for an *exact* calculation would always retain a pure state.

This type of interpretation of an asymmetry in time is not new. In chapter 3 we discussed how the introduction of coarse-graining leads to asymmetric entropy changes, provided considerations are restricted to intervals of time which are very small compared to the Poincaré cycle time for the system. Nevertheless, the exact microscopic behaviour of the (isolated) system is still reversible, and undirected in time. In quantum measurement theory, the coherence time of the apparatus wave function is analogous to the Poincaré time; in both cases there is a temporal asymmetry because of the limitations of human observations in real situations. The underlying reversibility of the total system is always present. Furthermore, the irreversibility which certainly always does enter in to the microscopic behaviour of object system I on measurement is closely similar to the classical microscopic irreversibility (increase in fine-grained entropy) which arises from random perturbations of the outside world intruding through the walls of the system container, and discussed in section 3.5. Just as the state vector in Hilbert space jumps abruptly in a non-unitary fashion because a measurement by an external

apparatus precludes a Hamiltonian (unitary) evolution, so the representativ
point in Γ space jumps about abruptly when outside perturbations preclud
a continuous trajectory described by a total Hamiltonain.

In both cases, consideration of the larger system restores the Hamiltonia
description, and the reversibility. In neither case does the irreversibility of th
subsystem make a significant contribution to the time asymmetry of th
system.

This analogy may be taken still further. By regarding the coupling of
closed system to the outside world as a continuous sequence of unrea
quantum measurements, the quantum $\bar{H}$ theorem may be allowed to res
entirely on the Klein lemma[14]. It follows from equation (6.61) that eve
though $\hat{\rho}$ is diagonal initially, off-diagonal elements gradually grow thereafter
and it is these that are responsible for the inequality of Klein's lemma. Th
continuous sequence of measurements has the effect of *diagonalizing* $\hat{\rho}$ agai
and again, by destroying the coherence and removing the interference terms
This is then somewhat analogous to the situation demanded by postulat
E of section 3.3 (see remark on line 3, p. 67). It will now be appreciated tha
the $\bar{H}$ theorem resting on the Klein lemma alone is the quantum analogue o
the 'external world' irreversibility discussed in section 3.5.

Having understood the nature of the quantum measurement irreversibilit
in this latter interpretation of quantum mechanics, it is instructive to inquir
after the origin. This clearly is connected with the macroscopic distinguish
ability of the measurement apparatus states, which in turn is a consequenc
of the internal structure of the apparatus. As emphasized by many authors[15]
an essential property of measuring equipment is that the initial condition o
the apparatus be *thermodynamically metastable*. Many real examples confirn
this—a cloud chamber is set up to be unstable to formation of water droplets
a photographic plate is unstable to chemical reactions in the presence o
light, etc. Metastability is an indispensible prerequisite for *amplification* of th
quantum signal which triggers the apparatus. As a result of this amplification
the apparatus proceeds fairly rapidly to a stable equilibrium condition, whic
is correlated with the initial state of I, and is macroscopically distinct fron
its neighbour states which are correlated with different states of I. Suc
amplification obviously involves a *coarse-grained* entropy increase (thoug
not fine-grained). Thus, although the *nature* of quantum measurement ir
reversibility depends on coherence times of macroscopic systems, the *origi*
of the irreversibility lies squarely in the domain of thermodynamics (as in th
Everett interpretation), which in turn is related (as described in chapters 3 an
4) to the formation of branch systems, and ultimately to the origin of th
universe, and the nature of gravity. Detailed analyses of the quantun
statistical mechanics in the measuring apparatus have been made by Daner
Prosperi and Loinger[16].

Once the thermodynamic character of the origin of quantum measurement irreversibility has been accepted, one may inquire into the status of the measurement process inside a totally isolated Boltzmann type box, where fluctuations occur, and entropy changes symmetrically in both time directions. It is indeed possible[17] to verify the consistency of the standard rules of quantum mechanics for measurements involving prediction in *both* time directions (recall the caution at the end of section 3.3).

Consider an experimenter A who prepares $\mathcal{N}$ systems in a state ϕ_n belonging to some complete set, and another experimenter B who at a later time in the conventional sense measures another quantity, the corresponding operator of which has the complete set of eigenfunctions $\{\psi_m\}$. The number of systems that B will find to be in the state ψ_m is $\mathcal{N}\ |\langle\psi_m \mid \phi_n\rangle|^2$ according to the postulates of quantum mechanics. On the other hand, if B is asked, on the basis of his measurement which reveals the system to be in some particular state, what was the result of A's earlier measurement, no definite answer could be given because the irreversibility of the measurement process has destroyed the necessary information. However, if the experiment is repeated, and A instead prepares his states *at random* with equal *a priori* probabilities, then a certain number of these systems will be found by B to be in a state ψ_m. If the question is now asked of B what fraction of these ψ_m systems were registered by A as being in a state ϕ_n the result is, of course, $\mathcal{N}\ |\langle\phi_n \mid \psi_m\rangle|^2$. Thus, there is complete symmetry with respect to the measurement process. In practice, the assumption by B of equal *a priori* probabilities for A's preparation is not justified, because the $\mathcal{N}$ systems would have been formed by interaction with the outside world as branch systems, a process which imposes non-random selections on the initial states ϕ_n. But in a totally permanently isolated system, equal *a priori* probabilities are justified, so that the asymmetry of quantum measurement disappears on long time scales, as expected. However, during a rare fluctuation, it would be necessary to use either the forward or backward prediction procedure described above, depending on the direction of entropy change at the time.

6.4 *T* violation in elementary particle processes

So far throughout this book it has been assumed that the nature of all forms of time asymmetric behaviour is rooted in the boundary conditions of the system (factlike rather than lawlike asymmetry). A closed permanently isolated system cannot show a permanent orientation in time direction in this case, as has been emphasized many times. Ultimately this fact is a consequence of the microscopic reversibility of the system constituents, as expressed through the classical equation (2.14) or the quantum equivalent (6.12).

This reversibility can be expressed as an invariance of the equations of motion for the system constituents under time reversal transformations, or T transformations as they are known in elementary particle physics (see section 1.6).

For many years it was felt that all the laws of physics ought to be T invariant. This is certainly the case for particles moving under the influence of electromagnetic and gravitational interactions, but the discovery of the strong and weak elementary particle interactions has called into question the whole subject of symmetry principles. In 1957, it was discovered[18] that the weak interaction (responsible for β decay) violated P invariance, i.e. invariance under *parity* transformations, which is the reflection of the spatial coordinate in the origin. It was still hoped that invariance under the combined operation of charge conjugation C (changing the sign of the charge) and P, might be preserved, but eventually even CP invariance had to be abandoned[19]. Now, there exists a well-known theorem which demonstrates that, under certain rather weak and plausible assumptions, the combined operation CPT must be invariant[20]. If CPT is to be retained, T must go. Indeed, this is found to be the case[21] in the decay of the neutral K meson

$$K^0 \to \pi^{\mp} + e^{\pm} + \bar{\nu}_e$$

or

$$\to \pi^{\mp} + \mu^{\pm} + \bar{\nu}_{\mu}.$$

This process therefore constitutes the only known example of where the *dynamical law* itself is oriented preferentially in time.

Although a T violating law of physics is of the greatest significance from a fundamental point of view, it is not clear that the properties of K mesons really have any relevance to the type of asymmetric processes that have been under discussion in this book. In particular, it has actually been demonstrated that it is not possible, by an appropriate use of K meson decay, to bring about an entropy decrease in an isolated system, i.e. the second law of thermodynamics emerges unscathed[22]. The essential feature of the demonstration is that *unitarity* is still preserved, and it is this requirement, rather than detailed microreversibility, which is necessary for the validity of the H theorem. It is possible to demonstrate via a sum rule that unitarity will impose an upper limit on the magnitude of the T violating part of the interaction which just corresponds to the limit beyond which entropy could be decreased. Of course, the presence of a T violating dynamical component can also be made to contribute to the entropy increase. In this respect it is on the same footing as the residual coupling of the system with the outside world, or the quantum measurement process. In all three cases there is a breakdown of microscopic reversibility, but the relaxation time of the system to coarse-grained equilibrium remains the dominant feature of time asymmetry.

6.5 Tachyons

A second assumption that has been made throughout this book is that all particles under discussion move along timelike world lines. It was shown in section 1.1 that such an assumption is necessary for the chronological ordering of a sequence of events to be preserved under Lorentz transformations.

It may be conjectured that there exists a class of particles that move along *spacelike* trajectories, i.e. travel faster than light. Such hypothetical particles are called *tachyons*[23]; there is no experimental evidence for their existence. It is usual to refer to ordinary massive particles as *tardyons*, and massless particles as *luxons*.

Special relativity requires that the energy E and momentum $\boldsymbol{p}$ of a particle form a four-vector p^μ with invariant length m_0^2:

$$p^\mu p_\mu = E^2 - \boldsymbol{p}^2 = m_0^2. \tag{6.77}$$

m_0 is known as the *rest mass* of the particle, and is a pure real quantity for all known particles. Consequently, the energy-momentum vector p^μ is timelike, and $E > |\boldsymbol{p}|$. When the particle is at rest $\boldsymbol{p} = 0$ and $E = m_0$ (positive root taken). At the other extreme, for very high momenta $E \simeq |\boldsymbol{p}|$ and the particle velocity approaches that of light. The relativistic mass m is related to m_0 by the equation

$$m = \frac{m_0}{(1 - v^2)^{\frac{1}{2}}} \tag{6.78}$$

for a particle moving with speed v. As $v \rightarrow 1$ (speed of light) $m \rightarrow \infty$. However, if m_0 were pure imaginary, p^μ would be spacelike and $|\boldsymbol{p}| < E$. This would not contradict common sense, provided $v > 1$, for then m would still be real in (6.78). The rest mass m_0 cannot be measured of course, as the particle is not allowed to be brought to rest.

The properties of tachyons are rather strange. For example, the most stable state (zero energy) is one of infinite velocity. More alarming still are the consequences of the breakdown of chronological ordering along the tachyon world line. If they are permitted to interact with tardyons, one could envisage a tachyon 'telephone line' with which two experimenters could communicate[24]. Suppose A sends a signal along the line to B. From the viewpoint of an observer in motion with respect to A and B, a situation can arise in which the direction of the signal is reversed. This need cause no trouble if we are willing to abandon the time order of cause and effect. Nevertheless, by contriving more complicated situations involving several experimenters equipped with devices for sending and receiving tachyons, paradoxical situations can occur[25].

These involve such unsavoury possibilities as the ability to signal one's own past. It is then possible to construct a machine which will only transmit a signal if it doesn't receive this same signal at a prior moment. In that case if the machine does receive the signal, it will not transmit it—an obvious contradiction.

These rather unpalatable properties of tachyons need not preclude their existence if it is supposed that the interaction between tachyons and tardyons is either non-existent, or random and uncontrollable[26].

6.6 Quantum action-at-a-distance electrodynamics

In section 5.8 a theory of electrodynamics was explained which is based on a direct interaction along the null cones between distant charged particles. This theory is found to be equivalent in all respects but one to Maxwell–Lorentz electrodynamics based on the interposition of an electromagnetic field as a mediator of influences between charged particles, provided the system was located inside a completely absorbing enclosure. The one difference between the two theories is that the action-at-a-distance electrodynamics does not possess divergent particle self-energies.

In quantum electrodynamics the situation is quite different. Self-energy effects may arise in new ways; for example, the emission and reabsorption of a virtual photon. However, the value of the self-energy now depends on the physical situation in which the particle finds itself. In particular, it is different for a free electron and one bound in a hydrogen atom. Although both quantities appear to be infinite in the conventional theory, this difference is in fact finite, though small. It has been measured to great accuracy[27], and agrees extremely well with refined theoretical estimates. Clearly, the action-at-a-distance theory cannot predict such an effect unless some sort of self-action is reintroduced. It turns out that this self-action leads inevitably to the divergences which the new theory sets out to avoid. There is no doubt that this considerably reduces the significance of the Wheeler–Feynman absorber theory of section 5.7, with all its cosmological implications.

The usual method of quantizing a classical theory consists of the construction of a Hamiltonian for the system, a choice of canonical coordinates and the imposition of canonical commutation rules. This procedure cannot be followed in an action-at-a-distance theory, owing to the non-local nature of the Lagrangian (see equation (5.83) which contains integrals over whole world lines) which prevents the construction of meaningful Hamiltonians[28]. It is possible to carry out a quantization in this case based on the action functional (5.84) by using Feynman's[29] sum over paths method; this has been done by Hoyle and Narlikar[30].

A more direct approach[31] is to work with the asymptotic S matrix, which is the usual formalism for quantum electrodynamics. This quantity may be introduced into quantum field theory in the following way. First an evolution operator analogous to $U(t_2, t_1)$ of equation (6.10) is defined. The Hamiltonian for interacting charged particles may be divided into two parts: H_0, the unperturbed part, and H_I which represents the effect of the (small) interaction between the particles

$$e^{-iH(t_2-t_1)} = e^{-i(H_0+H_I)(t_2-t_1)}.$$

It is convenient to perform a unitary transformation to the interaction representation, in which $U(t_2, t_1)$ is redefined as

$$U(t_2, t_1) = e^{iH_0t_2}e^{-iH(t_2-t_1)}e^{-iH_0t_1}$$

and obeys the differential equation

$$i\frac{\partial}{\partial t_2} U(t_2, t_1) = H_I(t)U(t_2, t_1) \tag{6.79}$$

so that the time dependence is here determined only by the interaction part of the Hamiltonian, H_I.

Equation (6.79) has the formal solution

$$U(t_2, t_1) = P \exp\left\{-i\int_{t_1}^{t_2} H_I(t)\, dt\right\}$$

which may be verified by expanding the exponential and substituting term by term into equation (6.79). P is an operator which places products of time-dependent operators in the order of their time label.

In a scattering event, the initial state of the system corresponds to widely separated particles whose interaction is negligible. Subsequently the particles approach closely and interact, after which they separate again. The scattering or S matrix is defined as the limit of the evolution operator as $t_2 \to \infty$ and $t_1 \to -\infty$

$$S = U(\infty, -\infty) = P \exp\left\{-i\int_{-\infty}^{\infty} H_I(t)\, dt\right\}.$$

The treatment may be made fully covariant by replacing $\int H_I(t)\, dt$ by $\int \mathscr{H}_I(x)\, dx$, where $\mathscr{H}_I$ is the interaction Hamiltonian density, and the integral dx is performed over all space–time. In the case of quantum electrodynamics, $\mathscr{H}_I$ is the same (to within a sign) as the interaction Langrangian density, $\mathscr{L}(x)$[32]. Therefore S becomes

$$S = P \exp\left\{i\int \mathscr{L}(x)\, dx\right\}. \tag{6.80}$$

In the conventional theory one takes

$$(x) = -j^{\mu}(x)A_{\mu}(x) \tag{6.81}$$

where j^{μ} is the electric current operator for the sources and A_{μ} is an *operator* which acts in the Hilbert space of the photons, which are treated as an independent mechanical system.

The scattering amplitude for any particular electrodynamic process may be obtained from the matrix element

$$\langle f|\, P \exp\left\{-i\int j^{\mu}(x)A_{\mu}(x)\, dx\right\} |i\rangle \tag{6.82}$$

with appropriate initial and final states $|i\rangle$ and $|f\rangle$ respectively. Of particular interest are the elements in which both $|i\rangle$ and $|f\rangle$ correspond to the photon vacuum, which we shall denote by $|0_{\gamma}\rangle$. This is the case where no real photons exist to propagate to infinity; only internal (virtual) photons are present. For example, the scattering of charged particles is included, but not the radiative emission of photons into free space.

For this particular choice of asymptotic states it is possible to prove the equality[33]

$$\langle 0_{\gamma}|\, P \exp\left\{-i\int j^{\mu}(x)A_{\mu}(x)\, dx\right\} |0_{\gamma}\rangle$$
$$= P \exp\left\{\tfrac{1}{2}i\iint j^{\mu}(x)D_F(x-x')j_{\mu}(x')\, dx\, dx'\right\}, \tag{6.83}$$

where D_F is the Feynman propagator for photons defined by

$$D_F(x) = \delta[(x)^2] + i\pi\frac{PP}{x^2} \tag{6.84}$$

PP denoting the principal part.

Notice that the right-hand side of (6.83) does not contain the *field* operators A_{μ}. Thus, in the absence of real photons in the initial and final states, electrodynamics may be described entirely in terms of a current–current interaction, reminiscent of action-at-a-distance. Indeed, the exponents on the left and right of (6.83) may be considered as the quantum mechanical analogues of the interaction terms in the action functionals (5.20) and (5.84) respectively. However, in the present case $D_F(x-x')$ is not the same as $\delta[(x-x')^2]$ in (5.84), but contains also an imaginary part (see equation (6.84)). This imaginary part represents the possibility of on-shell photon exchange. Thus, although the operator on the left of (6.83) is unitary on account of the Hermitean nature of the interaction Lagrangian density (6.81), the operator on the right is not unitary owing to the imaginary part of D_F.

A quantum version of the action-at-a-distance theory is obtained by deleting the imaginary part of D_F, and using the expression

$$\langle f|\, P \exp\left\{\frac{i}{2} \iint j^\mu(x)\, \delta[(x - x')^2] j_\mu(x')\, dx\, dx'\right\} |i\rangle \tag{6.85}$$

as the *fundamental* S matrix in place of (6.82). The operator in the exponent of (6.85) is time symmetric and Hermitean, although the expression is still not unitary owing to the presence of the P operator. Nevertheless, there exist submatrices of (6.85) which are unitary. One example of such a submatrix is where $|i\rangle$ and $|f\rangle$ are chosen to correspond to the ground state van der Waal's interaction, where on-shell photons are not allowed anyway by the conservation laws. Another example is a collection of currents that form a completely opaque cavity, in which the Wheeler–Feynman absorber processes remove the on-shell photon amplitudes by interference. In this case the expression (6.85) is equal to the right-hand side of (6.83).

In order to recover matrix elements corresponding to emission and absorption of real photons, it is necessary to divide the universe into two parts, the local system of interest and the rest of the completely absorbing universe. The expression in the exponent on the right of (6.83) may then be factorized into two parts, one of which describes the emission of a photon from the local system, the other its subsequent absorption in the distant universe. These terms turn out[34] to be *formally* equivalent to the expressions calculated using the S matrix (6.82) with the photon operators A_μ.

Finally, it will be noted that expression (6.85) describes the source currents in terms of the operators j^μ which may operate on states which include many indistinguishable particles, in contrast to the classical theory where the currents j were labelled by the classically distinguishable particles—see equation (5.84). This distinguishability enables the proviso $i \neq j$ to be made in the classical interaction, but in the quantum case this has no meaning because we are dealing with operators, not functions. Therefore, expression (6.85) automatically contains self-action terms. This enables the observable self-energy effects to be recovered in the action-at-a-distance formalism, but it leads also to identical divergences as in the conventional theory. However, the removal of these divergences may now be made by modifying the propagator D_F directly (perhaps from gravitational or quantized space–time considerations[35]) without reference to a dynamical theory of photons.

References

1. P. P. Bocchieri and A. Loinger, *Phys. Rev.*, **107**, 337, 1957.
2. R. C. Tolman, *The Principles of Statistical Mechanics*, Oxford University Press, London 1938, sections 83 and 84.

3. For a review, see R. Jancel, *Foundations of Classical and Quantum Statistical Mechanics*, Pergamon Press, Oxford, 1969, chapters 3 and 4.
4. R. C. Tolman, *The Principles of Statistical Mechanics*, Clarendon Press, Oxford, 1938, chapter 12. R. Jancel, *Foundations of Classical and Quantum Statistical Mechanics*, Pergamon Press, Oxford, 1969, section 6.4.1.
5. O. Klein, *Z. Phys.*, **72**, 767, 1931.
6. See for example, L. I. Schiff, *Quantum Mechanics*, 2nd edition, McGraw-Hill, New York, 1955, p. 58.
7. See for example N. Bohr, *Atomic Theory and the Description of Nature*, Cambridge University Press, Cambridge, 1934, 1961; *Atomic Physics and Human Knowledge*, Wiley, New York, 1958. For a more recent review, see R. J. Hall, Bohr's interpretation of quantum mechanics, *Amer. J. Phys.*, **33**, 624, 1965.
8. J. von Neumann, *Mathematical Foundations of Quantum Mechanics*, English translation, Princeton University Press, Princeton, 1955.
9. See for example J. M. Jauch, *Foundations of Quantum Mechanics*, Addison Wesley, Reading, Massachusetts, 1968, p. 179.
10. E. P. Wigner, *Amer. J. Phys.*, **31**, 6, 1963; Remarks on the mind-body question, in *The Scientist Speculates* (Ed. I. J. Good), Heineman, London, 1961 and Basic Books, New York, 1962. Also F. London and E. Bauer, *La Théorie de l'Observation en Mecanique Quantique*, Hermann, Paris, 1939.
11. H. Everett, *Rev. Mod. Phys.*, **29**, 454, 1957. J. A. Wheeler, *Rev. Mod. Phys.*, **29**, 463, 1957. B. S. de Witt, *Physics Today*, **23**, 30, 1970. B. S. de Witt and N. Graham (Eds.), *The Many-Worlds Interpretation of Quantum Mechanics*, Princeton University Press, Princeton, 1973.
12. D. Bohm, *Phys. Rev.*, **85**, 166, 1952; **85**, 180, 1952; **96**, 208, 1954; *Rev. Mod. Phys.*, **38**, 453, 1966.
13. K. Gottfried, *Quantum Mechanics*, Benjamin, New York, 1966, vol. 1, p. 186. A. Peres and N. Rosen, *Phys. Rev.* B, **135**, 1486, 1964.
14. M. Born, *Ann. Phys.*, **3**, 107, 1948; *Suppl. Nuovo Cimento*, **6**, 161, 1949. M. Born and H. S. Green, *Proc. Roy. Soc.* A, **192**, 166, 1948.
15. A. Daneri, A. Loinger and G. M. Prosperi, *Nucl. Phys.* A, **33**, 297, 1962. D. I. Blokhinstev, *The Philosophy of Quantum Mechanics*, Reidel Publishing Co, Dordrecht, Holland, 1968, p. 91.
16. A. Daneri, A. Loinger, G. M. Prosperi, *Nucl. Phys.* A, **33**, 297, 1962. See also *Nuovo Cimento*, **44** B, 119, 1966 and A. Loinger, *Nucl. Phys.* A, **108**, 245, 1968.
17. S. Watanable, *Rev. Mod. Phys.*, **27**, 179, 1955; *Prog. Theor. Phys. Suppl.* Extra Number, p. 135, 1965. Y. Aharonov, P. G. Bergmann and J. L. Lebowitz, *Phys. Rev.* B, **134**, 1410, 1961. R. H. Penfield, *Amer. J. Phys.*, **34**, 422, 1966.

18. C. S. Wu, E. Ambler, R. W. Hayward, D. D. Hoppes, R. P. Hudson, *Phys. Rev.*, **105,** 1413, 1957.
19. J. H. Christenson, J. W. Cronin, V. L. Fitch and R. Turlay, *Phys. Rev. Letters*, **13,** 138, 1964.
20. T. D. Lee, R. Oehme and C. N. Yang, *Phys. Rev.*, **106,** 340, 1957.
21. R. S. Casella, *Phys. Rev. Letters*, **21,** 1128, 1968; **22,** 554, 1969. Y. Achiman, *Lettere al Nuovo Cimento*, **2,** 301, 1969.
22. Lectures delivered at the Erice Summer School on high energy astrophysics and its relation to elementary particle physics, June 16–July 6 1972, by Y. Ne'eman.
23. G. Feinberg, *Phys. Rev.*, **159,** 1089, 1967. O. M. Bilaniuk and E. C. G. Sudarshan, *Physics Today*, **22,** 43, May 1969.
24. O. M. Bilaniuk and E. C. G. Sudarshan, *Physics Today*, **2,** 43, May 1969, p. 47.
25. F. A. E. Pirani, *Phys. Rev.* D, **1,** 3224, 1970.
26. In this connection see A. Peres and L. S. Schulman, *Int. J. Theor. Phys.*, **6,** 377, 1972.
27. W. E. Lamb and R. C. Retherford, *Phys. Rev.*, **72,** 241, 1947.
28. However, see E. H. Kerner, *J. Math. Phys.*, **3,** 35, 1962; **6,** 1218, 1965. R. N. Hill and E. H. Kerner, *Phys. Rev. Letters*, **17,** 1156, 1966. F. J. Kennedy, *J. Math. Phys.*, **10,** 1349, 1969 and references therein.
29. R. P. Feynman and A. R. Hibbs, *Quantum Mechanics and Path Integrals*, McGraw-Hill, New York, 1965.
30. F. Hoyle and J. V. Narlikar, *Ann. Phys.*, **54,** 207, 1969; **62,** 44, 1971.
31. P. C. W. Davies, *Proc. Cam. Phil. Soc.*, **68,** 751, 1970; *J. Phys.* A, **4,** 836, 1971; **5,** 1025, 1972.
32. N. N. Bogoliubov and D. V. Shirkov, *Introduction to the Theory of Quantised Fields*, Interscience, New York, 1959, pp. 206–227.
33. A. I. Akhiezer and V. B. Berestetskii, *Quantum Electrodynamics*, Interscience, New York, 1965, p. 302.
34. P. C. W. Davies, *J. Phys.* A, **5,** 1025, 1972, section 3.
35. F. Hoyle and J. V. Narlikar, *Ann. Phys.*, **62,** 44, 1971, section 5. See also A. Salam and J. Strathdee, *Lettere al Nuovo Cimento*, **4,** 101, 1971.

Further reading

1. R. C. Tolman, *The Principles of Statistical Mechanics*, Oxford University Press, London, 1938. R. Jancel, *Foundations of Classical and Quantum Statistical Mechanics*, Pergamon Press, Oxford, 1969. These books contain clear expositions of quantum statistical mechanics and the *H* theorems, as well as useful remarks on the measurement problem.

2. B. S. de Witt and R. N. Graham, Resource letter IQM–1 on the interpretation of quantum mechanics, *Amer. J. Phys.*, **39**, 724, 1971. This paper contains extensive references on interpretation and measurement problems in quantum mechanics. Some standard textbooks on quantum mechanics also include sections on the measurement problem. See for example D. Bohm, *Quantum Theory*, Prentice-Hall, New Jersey, 1951, part 6; K. Gottfried, *Quantum Mechanics*, vol. 1, Benjamin, New York, 1966, chapter 4; D. I. Blokhinstev, *The Philosophy of Quantum Mechanics*, Reidel Publishing Co, Dordrecht, Holland, 1968, chapter 13. In addition some of the more profound problems are discussed in *Quantum Theory and Beyond* (Ed. T. Bastin), Cambridge University Press, Cambridge, 1971.
3. Papers which make specific remarks about time asymmetry in quantum measurement that are not included in the references are H. D. Zeh, On the irreversibility of time and observation in quantum theory, in *Foundations of Quantum Mechanics*, Academic Press, New York, 1971
4. B. S. de Witt and N. Graham, *The Many-Worlds Interpretation of Quantum Mechanics*, Princeton University Press, Princeton, New Jersey, 1973.
5. It seems to have gone completely unnoticed that a quantization of Wheeler–Feynman electrodynamics was carried out as long ago as 1950 by G. Ludwig, Formulation of divergence-free electrodynamics, *Z. Naturforsch*, **5a,** 637, 1950. See also G. Süssman, Spontaneous light emission in unitary quantum electrodynamics, *Z. Phys.*, **131,** 629, 1952.
6. P. L. Csonka, Advanced effects in particle physics, *Phys. Rev.*, **180,** 1266, 1969.

7 Worlds Without End?

7.1 The heat death of the universe

Recalling the discussion of the Friedmann cosmological models in section 4.2, it is evident that conventional cosmology will permit two possible ends to the present condition of the universe. Either the whole universe will be returned to a very hot high-density phase once again, or it will continue to expand and cool for ever. In the laboratory, an isolated system soon reaches a state of equilibrium from which no perceptible change occurs over time scales of usual interest. In this condition (which is one of uniform temperature) no useful work may be extracted from the system, and the matter may be considered for all practical purposes to have reached its final state.

In the case of the whole universe the process relaxation times are very much longer than those of laboratory processes, but a straightforward application of the laws of thermodynamics leads to the inevitable conclusion that eventually, if the expansion continues, equilibrium will prevail throughout, and the entire universe will reach a final state from which no change will occur. This particular outcome is known as the *heat death* of the universe.

The precise nature of the heat death differs from the laboratory case. As stressed in chapter 4, the most significant thermodynamically irreversible processes from a terrestrial standpoint are those that are a consequence of the degradation of sunlight. Obviously, when the stellar nuclear fuels are exhausted, all starlight will cease (though new generations of stars may form before this happens). The end state of a star will be either a white dwarf, neutron star or black hole. Of these, only the black hole represents the cosmological final state of matter. Matter which is not in this condition is really metastable (section 4.6). Within the galaxies, a slow accretion of surrounding material by existing black holes would be expected. Interstellar collisions would eventually bring about the end of many stars in this fashion, while others would tend to evaporate away from the galaxy altogether. In the

Einstein–de Sitter model, where matter is only just 'unbound', this represents only a temporary reprieve, for any density perturbation would be expected to continue to grow by accretion, both of stars and intergalactic gas. In the $k = -1$ Friedmann model, the high expansion rate would rescue much of the uncollapsed matter from this canabalistic fate. (There may be an exceedingly small, though finite, chance that even these objects may quantum mechanically 'tunnel' into black hole oblivion—see also the end of section 4.6.)

The entropy of the intergalactic gas and radiation will also continue to increase by ionic recombination and photon absorption. Because of the red shift, all unabsorbed photon energies will eventually drop below the lowest binding energy of the intergalactic gas, after which absorption will only occur by unrecombined ions. The rate of cooling of the ions will depend on the precise balance between the heating due to photon absorption and the cosmological expansion. The rate of energy loss of ions at a temperature T_i by the expansion is proportional to $\dot{T}_i$. On the other hand the rate of gain of energy from a photon field of energy density ε_γ is proportional to $\varepsilon_\gamma \sigma$, which from equations (4.13) and (5.88) is seen to be proportional to $R^{-4}T_i^{-\frac{1}{2}}$. A balance will occur at a temperature $T_i \propto (\int R^{-4}\, dt)^{\frac{2}{3}} = R^{-\frac{8}{3}}t^{\frac{2}{3}}$ in the case that R changes as a power of t, as it does for the later stages of the Friedmann models. For example, the $k = -1$ model expands like $R \propto t$ eventually, so $T_i \propto R^{-2}$ and the ions cool at the same rate as the neutral matter. For the Einstein–de Sitter model $R \propto t^{\frac{2}{3}}$, so that $T_i \propto R^{-\frac{5}{3}}$, i.e. a rate intermediate between the radiation and neutral matter. Finally if $R \propto t^{\frac{2}{5}}$ we have $T_i \propto R^{-1}$ and the ions are in equilibrium with the radiation, cooling at the same rate. (In deriving this result the depletion of the photons has been ignored because the photon density in the universe is so enormously greater than the ion density.) Of course, this is just the condition for total absorption of the radiation found in section 5.9 another way. As discussed in that section, this situation can never arise because total recombination of the ions occurs anyway if R expands more slowly than $t^{\frac{1}{2}}$.

From this unsophisticated analysis, it appears that the *process rate* of all entropy increasing phenomena will eventually tend to zero, though the universe may never reach its maximum entropy state. In the $k = -1$ Friedmann model there will always be some uncollapsed matter and unrecombined ions; the radiation will not be totally absorbed, so that a temperature difference will always exist between the radiation and matter. A sufficiently resourceful intelligence could tap this apparently limitless store of useful energy, by equalizing temperatures and stimulating black hole cannabalism, perhaps extracting the energy in the form of gravitational radiation. In this sense, then, there is no true heat death in the $k = -1$ model. On the other hand, in the Einstein–de Sitter model, the heat death may well be complete.

Unhappy about this dismal choice of death by equilibrium or fiery collapse, many authors have proposed theories aimed at producing cosmic perpetuum mobiles. These will be examined in turn.

7.2 The steady state theory

One of the most controversial 'no heat death' theories is the steady state model, which at one time received sufficient acclaim to elevate it as a rival to the Friedmann models[1].

As the name implies, this theory avoids the irreversible death from equilibrium by means of a continual resuscitation in the form of injection of negative entropy into the universe. This comes about by the continual creation of matter, say in the form of hydrogen atoms spread uniformly throughout space. Creation occurs at a rate which just maintains the same average density (on a large scale) everywhere, in spite of the cosmological expansion. Such a rate of creation would be very low (about one atom per 10^{15} cm^3 per year) and quite undetectable. Nevertheless, this newly created matter would eventually accumulate in the empty spaces between the retreating galaxies, where it would condense into new galaxies, which would in turn pass through their usual entropy increasing activities to their own local heat death. However, the overall entropy per unit volume would remain constant.

The steady state model can be based on the so-called 'perfect cosmological principle' which asserts that the large scale features of the universe should remain unchanged with time, as well as with spatial location and orientation (cosmological principle). Thus, the new principle requires that the Hubble parameter $\dot{R}/R$ be constant, which implies that

$$R \propto e^{Ht}. \tag{7.1}$$

The universe clearly has no beginning and no end in this theory, so there is no big bang. The low entropy condition of matter at present cannot therefore be attributed to processes occurring in the early fireball, but follows instead as a consequence of the creation conditions (matter smoothed out and in the form of hydrogen). An attempt to construct a physical mechanism for the creation process was made by Hoyle[2], with the introduction of a negative energy creation or C field. The model may then be treated within the framework of general relativity, by solving the field equations (4.2) with a C field term in $T_{\mu\nu}$. There does indeed exist a solution which tends to a steady state condition for $t \rightarrow \pm\infty$ (see figure 7.1). The model is therefore time symmetric, the direction of thermodynamic processes in each branch being determined by the direction in time of the creation. Unfortunately, the C field cosmology only admits the advanced solutions of the absorber theory in the steady state limit[3].

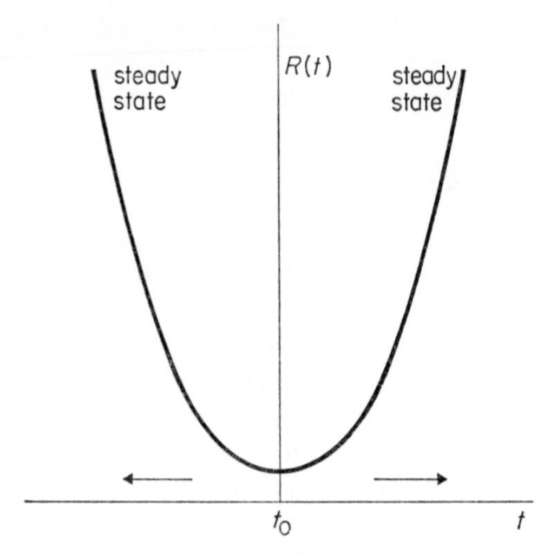

Figure 7.1 The C field cosmology. R is symmetric about t_0 and tends to a steady state condition ($R \propto e^{\pm Ht}$) for large $|t|$. On both sides of t_0 matter is created and entropy increases the time direction indicated by the arrows.

In recent years mounting observational evidence in astronomy, in particular the discovery of the cosmic microwave background radiation (which appears to demand an earlier dense state), has led to the almost complete abandonment of the steady-state theory.

7.3 The oscillating universe

This theory is based on the closed Friedmann model described in sections 4.2 and 4.4, which expands from a dense state to a maximum value of R after which it recontracts to a final dense state. The new feature to be discussed here is that this final dense state may in turn be followed by a renewed expansion, and so on, so that the universe oscillates forever between a maximum and minimum value of R (see figure 7.2).

It is not clear to what extent this model is consistent with general relativity which (at least in its unquantized form[4]) apparently demands that the minimum value of R must correspond to a singularity, $R = 0$, $\varepsilon = \infty$. If this is indeed so, it is not possible to continue the consideration of the physical condition of the universe through the extremities of each cycle. In spite of this, many authors[5] have speculated about a continually oscillating model, so for the purpose of this section we shall not take this objection as seriously as we perhaps should.

The way in which this model avoids the heat death is as follows. Each new cycle of expansion and contraction starts with a dense, hot fireball. The physics of this condition was discussed in chapter 4, where it was demonstrated how the origin of thermodynamic irreversibility lay in the changing constraints in the early stages. Consequently, *every* new cycle will emerge from the initial dense state with some negative entropy, and in the subsequent expansion and contraction the entropy of the cosmological fluid will increase from the various processes discussed in chapter 4. The dense state, which is assumed to be violent enough to destroy the stucture of the previous cycle, thus has the effect of 'starting the universe off again', a phenomenon which apparently continues indefinitely in both directions of time.

This result, of entropy increasing without limit in cycle after cycle, is in marked contrast with what would be expected from ordinary laboratory thermodynamics. Consider a homogeneous fluid undergoing a series of expansions and contractions in a local system[6]. The entropy increase is determined through the first law of thermodynamics (equation (2.2))

$$dS = \frac{1}{T}\,dE + \frac{p}{T}\,dV + \sum_i \frac{\partial S}{\partial n_i}\,dn_i \tag{7.2}$$

where the energy E, volume V and number of moles n_i of the different fluid components are the independent variables of the system. If the system is isolated and adiabatic, both dE and $p\,dV$ are zero on account of the conservation of energy and the isolation respectively. In this case the only remaining possibility for entropy to increase comes from the redistribution of concentrations of the different fluid constituents, as described by the last term of (7.2). This cannot continue indefinitely, because for a given energy and volume there is a maximum value of the entropy corresponding to chemical equilibrium between the different fluid components. When this condition is attained there can be no further change in the fluid. On the other hand, if the system is not adiabatic, but is coupled to an external energy supply (for example, a piston and cylinder arrangement being driven by a motor), the entropy will continue to rise until the external energy source is exhausted.

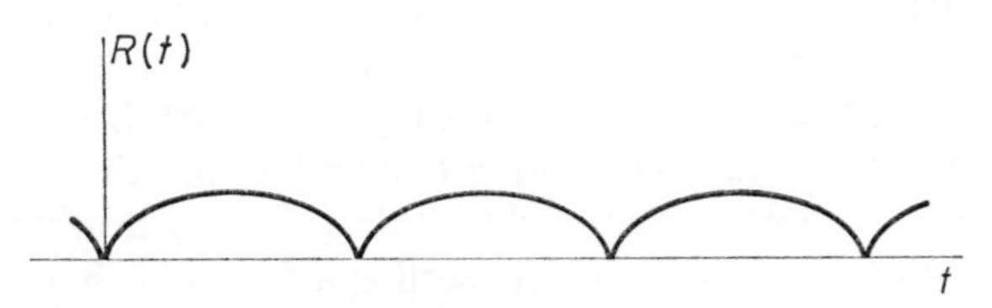

Figure 7.2 The universe oscillates forever between a maximum value of R, and a minimum at which known physics breaks down.

This comes about due to the ever present lag of the expanding and contracting fluid behind equilibrium conditions.

In the cosmological case there is an equation analogous to (7.2) for the entropy in a comoving volume, as measured by a local observer performing experiments in his immediate neighbourhood in the comoving frame

$$dS = \frac{1}{T} d(\varepsilon R^3) + \frac{p}{T} dR^3 + \sum_i \frac{\partial S}{\partial n_i} dn_i. \tag{7.3}$$

The first two terms on the right of equation (7.3) vanish *together* on account of equation (4.6)

$$d(\varepsilon R^3) + p\, dR^3 = 0. \tag{7.4}$$

Also in the cosmological case, the entropy of the fluid cannot increase as a direct result of the expansion and contraction. However, the presence of the last term in (7.4) again opens up the possibility that a multicomponent fluid might undergo a readjustment of concentrations between the components so as to achieve an entropy increase. In practice, a very important process of this kind is the emission of starlight, which increases the concentration of the radiation field at the expense of the matter.

In the discussion of the one-component fluid-filled oscillating models given in section 4.4, the scale factor $R(t)$ was a precisely symmetric function about the maximum value; these models were exactly reversible. In the present case, where the cosmological fluid may undergo an increase in entropy through multicomponent readjustments, the reversibility of the entire system is destroyed. There is no longer complete symmetry between the expanding and contracting phases of the cycle, though realistically the departures from symmetry are small.

To understand how these departures come about, we appeal to the well-known failure of the principle of energy conservation in general relativity. Even an isolated system need not have constant energy in the usual sense. Indeed, inspection of equation (7.4) reveals that, owing to the presence of the second (positive) pressure term, the energy density ε of every element of fluid is in general decreasing on expansion and increasing on contraction. If, therefore, p takes different values during the expansion from the subsequent contraction, the element of fluid will be returned to its former volume with increased energy. Such a pressure difference might occur if an adjustment of concentrations was taking place with a finite relaxation time, so that there was always a lag behind equilibrium conditions. This is clearly the case with starlight emission which, as shown in section 4.5, lags behind the expansion time scale by many orders of magnitude. Quite simply, the pressure of starlight is greater in the contracting phase because the stars have been shining

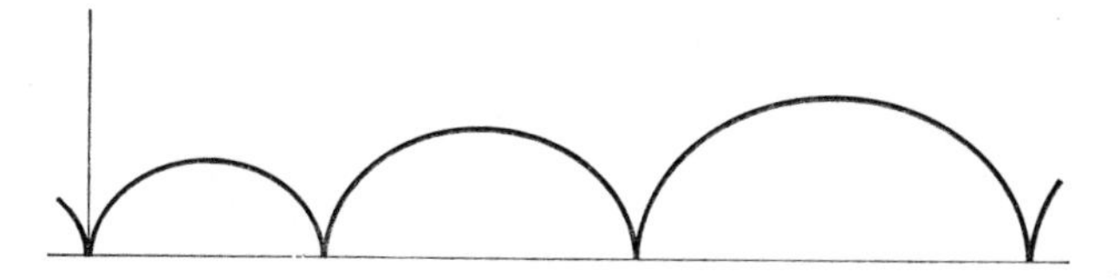

Figure 7.3 Irreversible processes in each cycle cause a gradual growth of amplitude.

longer, so there is more starlight around. Consequently, the energy of a co-moving region of the universe may be expected to have increased slightly at the same volume after one cycle. Furthermore, the appearance of this increased energy density can be attributed to a corresponding decrease in the *gravitational potential energy* of the fluid, which plays a role analogous to the external energy source of the laboratory piston and cylinder experiment. However, unlike the laboratory case, the gravitational energy of the system is limitless on account of the singular nature of the terminal points of infinite compression. The continued increase in the entropy of the cosmological fluid is thereby explained. (The reader is reminded of the discussion of black holes given in section 4.6 in which it was pointed out how these objects acquire infinite entropy at the expense of gravitational potential energy. The whole universe can be considered as a black hole in the closed Friedmann model.)

Curiously, rather than the irreversible processes tending to damp out the oscillations, as would be expected from classical considerations, they actually continue to grow (see figure 7.3). This result may be deduced from inspection of equation (4.4) which shows that, for a given value of R, a larger value of ε implies a larger value of $\dot{R}$.

It might be wondered how this progressive growth of oscillations, and the general departure from symmetry during each cycle, is consistent with the uncontestable invariance of Einstein's field equations under time reversal. As mentioned in section 5.1 in connection with the wave equation, it is necessary to reverse the *sources* of the field also if time symmetry of the solutions is to be recovered. In the field equations (4.2) of general relativity, the source term is the stress-energy-momentum tensor $T_{\mu\nu}$, which is of course a *macroscopic* quantity, not itself normally subject to reversible behaviour. A truly symmetric reversible cycle could only be achieved by arranging for the *microscopic* reversal (in the Loschmidt sense) of all the particles and waves in the universe at the point of maximum expansion. This very special situation has indeed been proposed by some authors and will be examined in the next section.

7.4 Time symmetric cosmologies

One obvious way of avoiding the heat death is to contrive a cosmological model in which there is complete time symmetry (in the classical sense now), so that an approach to equilibrium is prevented by a 'regenerating' fluctuation to low entropy values. As mentioned in section 4.7, Boltzmann himself made such a suggestion, by regarding the present low entropy condition as a gigantic fluctuation of the whole universe from an equilibrium condition. In a static Newtonian universe such fluctuations would occur infinitely often, over time scales of immense duration (the Poincaré time for the universe is measured in at least $10^{10^{80}}$ Hubble times!). In the ever expanding cosmological models, fluctuations on scales of galactic dimensions would be expected in spite of the dissipation due to expansion. One is then faced with the amusing problem of why we are obviously living in the *first* low entropy condition of our galaxy, when there are an infinite number of other low entropy occasions, but in which the universe would be so expanded (after the enormous duration) that the other galaxies would be quite invisible to us. A solution to this problem is provided by the existence of black holes in general relativity (section 4.6) which do not permit fluctuations to low entropy states (except possibly in a quantum theory). If all galactic matter is eventually accreted by black holes (which might occur in the Einstein–de Sitter model—see section 7.1) then this would guarantee a permanent heat death.

In the oscillating model, troublesome Poincaré fluctuations do not have time to occur unless we make the duration of the cycle incredibly long. It would then be possible to identify the minima of two adjacent entropy fluctuations with the singular terminal points of an oscillation. Although elegant, this would be quite without physical justification, for one must not confuse the symmetry of the cosmological model as governed by the equations of general relativity with the symmetry of a closed system of interacting particles as governed by the principles of statistical mechanics—two quite disjoint aspects of physics.

To emphasize this last remark, consider a model universe homogeneously filled with pressure free dust, exactly at rest with respect to the comoving frame. In this case, the reversal of the cosmological expansion *would* bring about a Loschmidt type reversal of particle velocities with respect to a comoving observer looking at distant matter. As for the local matter, this is at rest in the comoving frame anyway, and of course, as we know from the discussion of section 4.4, this is a thermodynamically reversible universe with no entropy change. In contrast to this, suppose that the motionless dust is replaced with a real gas having local inhomogeneities and thermal motion in the comoving frame. An observer performing experiments on small quantities of gas in his neighbourhood would deduce the existence of fluctuations, Poincaré cycles and the second law of thermodynamics. He would conclude that there is *no reason* to suppose that a remarkable entropy decreasing fluctuation in his local neighbourhood would just happen to be initiated at the moment of maximum cosmological expansion, itself a quite unremarkable occurrence, as explained at the end of section 4.5.

In view of these remarks, it is evident that the still stronger correlation between the cosmological motion and thermodynamics proposed some time ago by Gold[7] should be regarded with some scepticism. In Gold's model, the cosmological fluid undergoes a steady entropy decrease in the contracting phase of the oscillating cycle, not as a result of a Poincaré fluctuation, but as a consequence of the diminishing volume of space. In this way the local thermodynamic processes are supposed to be strongly coupled to cosmology. This can only occur by choosing the initial state of the universe to belong to a very special class containing strong correlations, in itself not an objectionable idea for these correlations would be quite unobservable. (This is obviously closely similar to the favourite idea of Bohm that there is 'folded-up order' in the universe about which we are no more aware than the order 'folded-up' in a hologram.) A neat way of selecting such a class of initial states is to impose *both* initial and final boundary conditions (of low entropy) on the system. Simple models based on the Ehrenfest urn experiment with symmetric boundary conditions have been discussed by Cocke[8].

To provide supportive evidence for his proposal, Gold appeals to the notion that the extreme thermodynamic disequilibrium between the stars and empty space is due to the fact that the universe is in a state of expansion, acting as a cold sink of energy. The emission of starlight is then supposed to be *due* to the cosmological motion, from which it is but a small step to conjecture that all thermodynamic disequilibria are rooted in this same cause. However, this argument seems to overlook the fact that the emission of starlight is an *irreversible* process, although the cosmological expansion is, in the homogeneous approximation, reversible. On p. 96 a careful distinction was drawn between the flow of radiation away from a hot star into the cold surrounding space, and the flow of radiation out of a fixed volume of space as the universe expands. Thus, a recontraction of the universe would certainly restore (reversibly) the temperature of the cosmic background radiation, but will not in general return the starlight to the stars (which are highly localized centres of emission) against their temperature gradients. These gradients still maintain the surface temperatures of the stars above the background temperature until well into the recontracting phase. The hypothesis of such a contrived situation as Gold has suggested must be supported instead by arguments referring to the special initial conditions necessary to bring it about; it does not receive support from the nature of starlight emission. In section 4.5

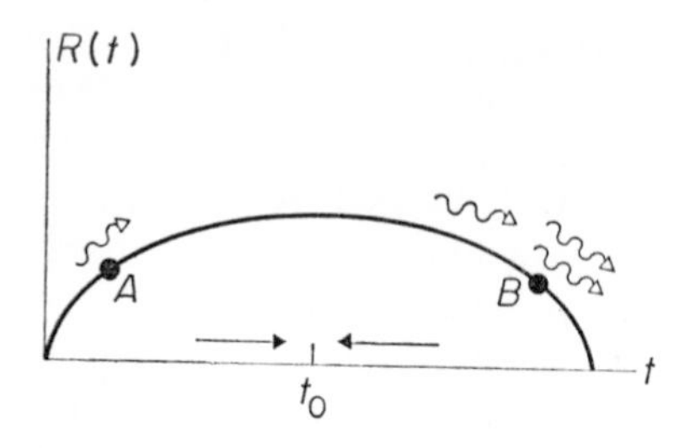

Figure 7.4 Gold's time symmetric cosmology. The arrows indicate the direction of entropy increase. If a photon emitted at A penetrates to B in the 'other half', it would not be absorbed but multiplied.

it was strongly emphasized that the emission of starlight was almost completely independent of the expansion of the universe and that, moreover, the darkness of the night sky can equally well be attributed to the finite age of the universe.

Further difficulties with Gold's model emerge from more detailed considerations of the 'switch-over' that occurs at the point of maximum expansion. As the universe is not even supposed to have reached thermodynamic equilibrium at that time, there will still exist diverging spherical wavefronts of light from hot stars, which will continue as diverging waves after the switch (assuming Maxwell's equations are uneffected by the change). This is merely an expression of the frequently mentioned fact that the distant galaxies would remain red shifted after the contraction had begun. A complete time symmetry about the point of maximum expansion would require an instantaneous changeover from diverging to converging waves (or having a mixture of both). It does not seem to be possible to contrive this in any plausible way.

If retarded waves from the 'first' half of the Gold universe are allowed to propagate into the 'second' half, and vice versa, then one runs into the energy escalation problems encountered in section 5.6, and which always seem to arise when one has a mixture of advanced radiation with entropy-increasing, charged matter. Any photons that get across the switch-over unabsorbed will find when they encounter matter that the prevailing thermodynamic processes are such as to produce *anti-damping* with respect to the other half of the cycle (see figure 7.4). If a light wave were to encounter the surface of a cold object, it would *multiply* in energy exponentially instead of diminish. For a metallic substance of only a thousandth of a centimetre in thickness, the skin depth is such that the resulting avalanche of photons would be greater than the total number in the universe! Consistency problems of this sort are bound to arise when oppositely directed regions of the universe are causally coupled together.

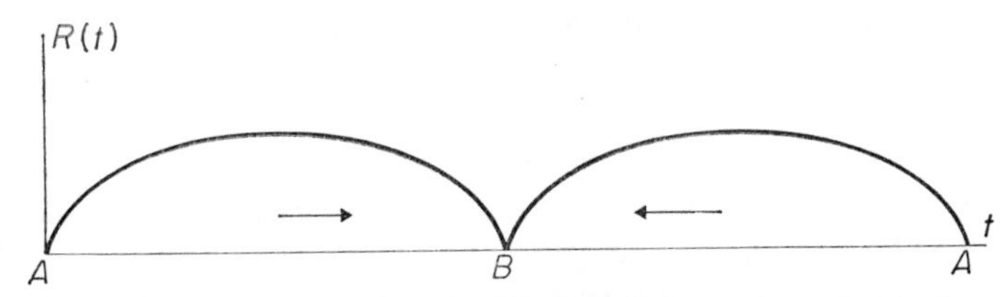

Figure 7.5 Closed time—imagine the diagram being wrapped round a cylinder so that the two *A*s coincide. The arrows indicate the direction of entropy increase, etc, which is opposite in the two 'halves'.

7.5 Closed time

Throughout this book there has been a tacit assumption that the *topology* of time is that of an infinite or semi-infinite straight line. However, speculation concerning the possible time symmetry of the whole universe would appear to be more appropriate in the context of closed time, having the topology of a circle. It may be pointed out immediately that this does not imply a periodic or recurrent world in which people will re-experience events in endless repetition*, for this is to invoke the dubious subject of the 'flow' of time.

The first major change in the structure of time attendant on the new topology is the abolition of the concepts 'before', 'after' and 'between', except for restricted regions of the time axis. Secondly, we are no longer free to 'select' initial conditions in the universe, and to follow their consequences for all eternity. Instead, there is a restriction to self-consistent cycles of reasoning, similar to those encountered in connection with the absorber theory of electrodynamics. Clearly these cycles must be symmetric in time.

The author has described[9] a model universe of this variety which consists of *two* cycles of expansion and recontraction (see figure 7.5). The direction of temporally asymmetric processes, such as entropy increase, is opposite in the two cycles, but unlike the Gold model, remains the same throughout each cycle. The points of contact between the two regions of opposite asymmetry are the dense states between the cycles. These are the natural points for the switch-over because the cosmological fluid reaches thermodynamic equilibrium there. Indeed, the minimum value of R, unlike the maximum, is certainly a very special condition of the world, and is often considered[10] to be the place where 'two halves' of the universe might be joined. (The foregoing is true even for local gravitational collapse. If a black hole can survive the singularity, it clearly cannot emerge into the same universe, for then an asymptotic observer would see it 'come out' of the horizon before it went in, which takes an infinite time.)

* If that were so, the author would be forced to agree with Eddington who wrote 'It seems rather stupid to keep doing the same thing over and over again.'

It might be wondered whether an observer in such a universe could tell in which half he resided. We would expect the operation of the observer's brain, memory and perception to be determined by thermodynamic factors. If it is assumed that the subjective notion of a flow of time is also rooted in the processes which occur in the brain, it seems that a casual observation would not permit one cycle to be distinguished from the other. A more ambitious individual would discover that the K^0 meson carries a time orientation which could be either parallel or antiparallel to the thermodynamic asymmetry, thus apparently providing a distinction between the two 'halves' of the universe. However, it has been shown by Ne'eman[11] that, provided *CPT* is valid, an interchange of the definition of matter and antimatter would remove this distinction. Expressed another way, the distinction disappears if we let matter change to antimatter at the switch-over[12].

An interesting result of the two cycle model follows if conservation of baryon number is allowed through the minima. Consider what happens to the starlight emitted during one of the cycles. As it passes into the hot dense state at the end, it will become thermalized, and emerge in the other cycle as a blackbody background. By symmetry, the starlight from the latter cycle will appear as a blackbody background in the former.

It is a straightforward matter to estimate the (constant) ratio of thermal photons to baryons in either cycle as a result of this. An upper limit to the energy per baryon emitted in the former cycle as starlight is obtained by supposing that *every* proton in the universe is 'cooked' in a star, releasing about 1% of its rest mass as starlight. The accumulated energy density will therefore be $10^{-2}\,\rho m$ for protons of number density ρ and mass m. It is an elementary result of radiation theory that the number density ρ_γ of blackbody photons is equal (to within a small numerical factor) to the $\frac{3}{4}$ power of the energy density in our units. Consequently $\rho_\gamma \simeq (10^{-6}m^3\rho^3)^{\frac{1}{4}}$, so that the photon/baryon ratio in the latter cycle is

$$\frac{\rho_\gamma}{\rho} \simeq (10^{-6}m^3/\rho)^{\frac{1}{4}}. \tag{7.5}$$

The value of the right-hand side of (7.5) depends on the value chosen for the matter density ρ, which is, of course, a function of epoch. This is an expression of the fact that no consideration has been made of the red shift of the starlight, which has the effect of weighting the contribution of photons emitted near the region of maximum expansion. As the closed Friedmann model spends most of its time in this region (see figure 4.1), the average value of ρ_γ over a whole cycle is unlikely to differ much from its present observed value. Calling the total number of particles inside the Hubble radius N, the proton density is of order NH^3, so that we obtain from (7.5)

$$\frac{\rho_\gamma}{\rho} \simeq \left[10^{-6}N^{-1}\left(\frac{m}{H}\right)^3\right]^{\frac{1}{4}}.$$

The ratio $\left(\frac{m}{H}\right)^3$ is just the total number of fermions that can 'fit' into the universe according to the Pauli principle. There is a well-known[13] coincidence that m/H is about 100 times $N^{\frac{1}{2}}$, i.e. about 10^{42}, so that

$$\frac{\rho_\gamma}{\rho} \simeq N^{\frac{1}{8}} \simeq 10^{10}. \tag{7.6}$$

The relationship (7.6) is not new, but has been noted[14] as an unexplained 'coincidence'. The result is an upper limit on the ratio ρ_γ/ρ, but the author has given another estimate[9] based on actual astronomical observations, which is only a little less than (7.6), about 10^8. Expressed in terms of the temperature of the black body background at our own epoch, this estimate turns out to be somewhere in excess of 2·4 K, which should be compared with the observed value of 2·7 K.

7.6 Conclusion

It is a remarkable fact that all the important aspects of time asymmetry encountered in the different major topics of physical science may be traced back to the creation or end of the universe.

Philosophers will distinguish two prevailing schools of thought regarding the creation event (setting aside for the moment the models which do not have one). Traditionally, the present condition of the universe has been regarded as a highly specific one, requiring the creation to have been of a very particular nature to ensure the appropriate subsequent structure. In recent years, developments in modern cosmology have had a profound impact on this question. Many people now incline to the opposite point of view, i.e. that the present state of the cosmos is a *typical* one, which would almost certainly result from a very wide range of initial conditions. As a conclusion to this book, it is interesting to see how the subject of time asymmetry has provided material for both these points of view.

The less exotic considerations of the previous chapters are consistent with the latter picture. To understand this, return for a moment to the subject of the formation of branch systems (section 3.4). Among the set of all possible initial conditions, the overwhelming majority correspond to situations in which the entropy will subsequently increase. A tiny fraction will correspond to those very special particle motions for which there will follow a temporary entropy decrease, while the set which gives rise to a permanently low entropy condition has a measure zero (see p. 48). The time asymmetry actually observed in a single branch system is therefore consistent with the initial conditions being chosen from this set at random. The parallel direction of entropy increase in vast numbers of branch systems is excellent *evidence* for this randomness.

In the cosmological case, an observation cannot be used to make a statement of a statistical nature about the entire universe, if the universe is unique, i.e. does not belong to an ensemble. Instead it may be concluded that a very wide range of initial conditions are consistent with the observed time asymmetry, so that the universe behaves *as if* it has been set up in a random fashion. That is to say, we do not need to look for something *special* about the world to account for its time asymmetry. Rather, we should have to look for something special if there were no asymmetry! For time asymmetry would appear to be a general feature of all possible cosmological models, except perhaps for a set of measure zero. (For example, a model in which the cosmological material consisted entirely of point particles exactly at rest in the comoving frame.)

In addition to this, there are reasons why we might actually *expect* the initial condition of the universe to be a 'random' one. First, there is no reason to suppose that the boundary conditions on the large scale motion of the universe, which determines the way in which it expands from the initial singularity at $t = 0$, should be matched with the boundary conditions on the individual particles of the cosmological material, i.e. there is no reason to suppose that the universe will start to expand out of a singularity at just the right moment to correspond to an overwhelmingly unlikely entropy decreasing fluctuation (a similar remark was also made on p. 193). Secondly, the existence of a particle horizon means that at sufficiently early epochs arbitrarily small regions of the universe were causally isolated (this remains true even if there was an imaginary mass component in the cosmological fluid)[15]. We would therefore not expect that particles in widely different parts of the universe would be initially correlated. Of course, the existence of the particle horizon in the first place depends upon the retarded nature of interactions. In conventional field theory this implies the non-existence of 'incoming' (i.e. correlated) photons, gravitons and neutrinos at $t = 0$, so that we have only removed the arbitrariness of one set of initial conditions by employing another equally arbitrary set. In the absorber theory the argument is a stronger one because the retarded nature of interactions itself demands the future thermodynamic irreversibility of the matter in the universe, in order that the radiation is absorbed. As usual in the Wheeler–Feynman theory there is a self-consistency requirement: the retarded radiation demands a 'random' initial state for the matter in order that it provide the appropriate thermodynamic conditions for absorption to occur in the far future, and this initial state depends in turn on the retardation. This is most satisfactory, for it is probable that any other condition for the universe (except of course the completely time reversed situation with decreasing entropy and advanced potentials) would not be self-consistent in this way (see also remarks on p. 130). Actually, the old controversy[16] about whether thermodynamic irreversibility implies retardation, or vice versa, has been resurrected here. Some authors[17] maintain the equivalence of both viewpoints.

Thirdly, there are in fact certain senses in which the universe may be considered to be a member of an ensemble. For example, in the continually oscillating model of section 7.3, the big bang does not represent a creation event anyway. Instead, there are an infinite chain of expanding and recontracting cycles, separated by dense states. These dense adjoining phases could be regarded as *reshuffling* the contents of the universe, so starting off each new cycle in a random manner. Another example is provided by the Everett interpretation of quantum mechanics (section 6.3) where the single world of our experience is supplemented by a whole ensemble in which all quantum alternatives are realized simultaneously. In this situation, many possible initial states may occur, the overwhelming majority in general leading to entropy increasing worlds. In any case, only those quantum worlds compatible with life will be observed.

In spite of the plausibility of an unremarkable origin of the universe, in which the expansion sets off without regard for correlated microscopic behaviour, there appears to be a paradox in the case of those models which recontract to a final singular state. Going in one direction of time, the universe expands and contracts with entropy increasing all the while, because of the lag of the cosmological contents behind the equilibrium conditions. On the other hand, going in the opposite direction of time, we start out from the same sort of singular condition, only now the microscopic behaviour of the cosmological material is strongly correlated together, and to the global motion, so that as the universe expands and contracts, the microscopic motions collectively *anticipate* the global motion, so contriving to decrease the entropy.

To illustrate these remarks further, return once again to the analogy of the recontracting universe with a laboratory gas-filled piston and cylinder coupled up to an external energy supply. So long as the energy supply lasts, the entropy of the gas will increase (with overwhelming probability) because of the lag of the gas motion behind the piston motion as explained in section 7.3. However, when the energy supply is exhausted, the system returns to time symmetric behaviour, for the reason that a combined fluctuation of the gas and piston plus energy supply would equally decrease and increase the entropy. Thus, with an available *external* energy source, the direction of temporal asymmetry of the outside world is imposed. (This is assuming that we don't wait for very rare entropy decreasing fluctuation before setting the mechanism off.)

Now in the cosmological case there is no external world which can impose a time asymmetry on the expanding and contracting cosmos. The energy source is in this case the gravitational self-energy of the universe which is inexhaustible. It therefore appears that the asymmetry between the two ends of a cycle of expansion and contraction is completely arbitrary, and must be accepted as a fact of nature. We simply *define* the past to be the temporal direction of the singular state at which there are no correlations between the microscopic and macroscopic motions of the world.

In section 7.4, it has been seen how some authors are unsatisfied with this arbitrariness, and have adopted instead the former of the two philosophical viewpoints mentioned at the beginning of this section. They propose that the universe was in fact created in a very special way, with initial conditions on the particle motions to just correspond to a time symmetric cosmos, in which the universe returns eventually to a low entropy state. The requirement of time symmetry then imposes severe constraints on the structure of the universe, so severe in fact that such a model seems to be incompatible with the observed universe if one single cycle of expansion and contraction is contemplated (of course in this case which end of the cycle is called the 'creation' is a matter of indifference). However, if one is prepared to accept the still more exotic possibility of topologically closed time then good agreement with the observed universe is obtained.

It seems that we have reached the limits at which *physics* can supply useful information regarding the *origin* of time asymmetry. The remaining problems in this respect appear to be philosophical rather than physical, and centre on questions such as what is the meaning of an ensemble of universes, how a single space–time can possess an overall asymmetry from within and whether the notion of closed time presents intolerable problems concerning matters of freewill. There is no doubt that these topics and others will continue to keep the philosophers busy for a long time to come.

References

1. H. Bondi and T. Gold, *Mon. Not. Roy. Astr. Soc.*, **108,** 252, 1948.
2. F. Hoyle, *Mon. Not. Roy. Astr. Soc.*, **108,** 372, 1948; **109,** 365, 1949.
3. P. C. W. Davies, *J. Phys.* A, **5,** 1722, 1972, section 8.
4. The possibility that a future quantized theory might avoid the singularity has been discussed in the context of the Friedmann cosmological models by B. S. de Witt, *Phys. Rev.*, **160,** 1113, 1967.
5. R. C. Tolman, *Relativity Thermodynamics and Cosmology*, Clarendon Press, Oxford, 1934, section 174. J. D. North, *The Measure of the Universe*, Clarendon Press, Oxford, 1965, chapter 18. W. Bonner, *The Mystery of the Expanding Universe*, Macmillan, New York, 1964 and Eyre and Spottiswoode, London, 1965, p. 123.
6. The argument presented here follows closely the discussion given in R. C.

Tolman, *Relativity Thermodynamics and Cosmology*, Clarendon Press, Oxford, 1934, pp. 135, 327–30. See also H. Zanstra, *Proc. Kon. Ned. Acad. Wetens.*, **60**, 285, 298, 1957.
7. T. Gold, The arrow of time, in 11th International Solvay Congress, *La Structure et l'Evolution de l'Universe*, Edition Stoops, Brussels, 1958; *Amer. J. Phys.*, **30**, 403 1962; Arrow of time, in *Recent Developments in General Relativity*, Pergamon–Macmillan, New York and Warsaw, 1962. In spite of its questionable physical basis, this imaginative theory has received considerable support. See for example H. Bondi, *The Observatory*, **82**, 133, 1962; D. L. Schumacher, *Proc. Cam. Phil. Soc.*, **60**, 575, 1964; M. Gell-Mann comments in *Proceedings of the Temple University Panel on Elementary Particles and Relativistic Astrophysics.* Y. Ne'eman, *Int. J. Theor. Phys.*, **3**, 1, 1970. P. T. Landsberg, *Studium Generale*, **23**, 1108, 1970, reprinted in *The Nature of Time* (Ed. J. T. Fraser, F. C. Haber and G. K. Muller), Springer, Berlin, 1972.
8. W. J. Cocke, *Phys. Rev.*, **160**, 1165, 1967. A paper by H. Schmidt, Model of an oscillating cosmos which rejuvenates during contraction, *J. Math. Phys.*, **7**, 495, 1966, seems to have been overlooked.
9. P. C. W. Davies, *Nature Physical Science*, **240**, 3, 1972.
10. See for example R. Penrose, Structure of space–time, in *Batelle Rencontres* (Ed. C. M. de Witt and J. A. Wheeler), Benjamin, New York, 1968, p. 222. F. Hoyle and J. V. Narlikar, *Mon. Not. Roy. Astr. Soc.*, **155**, 305, 1971.
11. Y. Ne'eman, *Int. J. Theor. Phys.*, **3**, 1, 1970. See also A. Aharony and Y. Ne'eman, *Int. J. Theor. Phys.*, **3**, 437, 1970.
12. M. G. Albrow, *Nature Physical Science*, **241**, 56, 1973.
13. See for example D. W. Sciama, *Modern Cosmology*, Cambridge University Press, Cambridge, 1971, p. 124.
14. E. R. Harrison, *Physics Today*, **25**, 30 December 1972.
15. P. C. W. Davies, Some cosmological consequences of imaginary mass (to be published).
16. A. Einstein, *Phys. Z.*, **10**, 185, 323, 1909; A. Einstein and W. Ritz, *Phys. Z.*, **10**, 605, 1909.
17. O. Costa de Beauregard, Irreversibility problems, in *Logic, Methodology and Philosophy of Science* (Ed. Y. Bar-Hillel), North Holland, Amsterdam, 1965, p. 326.

Further reading

1. For details concerning the 'end' of the universe, see M. J. Rees, The collapse of the universe: an eschatological study, *The Observatory*, **89**,

193, 1969 for the closed Friedmann model, and P. C. W. Davies, The thermal future of the universe, *Mon. Not. Roy. Astr. Soc.*, **161,** 1, 1973 for the ever expanding models.

2. Further investigation of the Gold model has been made by L. S. Schulman, Correlating arrows of time, *Phys. Rev.* D, **7,** 2868, 1973.
3. Speculation and references concerning the concept of cyclic or closed time may be found in J. D. North, *The Measure of the Universe*, Clarendon Press, Oxford, 1965, chapter 18; A. Grünbaum, *Philosophical Problems of Space and Time*, Knopf, New York, 1963.
4. A different sort of 'two part' universe with oppositely directed time sense was suggested by F. R. Stannard, Symmetry of the time axis, *Nature*, **211,** 693, 1966.

Index

Numbers in italics refer to sections